Critical Risk Research

Critical Risk Research

Practices, Politics and Ethics

EDITED BY

Matthew Kearnes

Francisco Klauser

Stuart Lane

WILEY-BLACKWELL

A John Wiley & Sons, Ltd., Publication

Wiley-Blackwell is an imprint of John Wiley & Sons, formed by the merger of Wiley's global Scientific, Technical and Medical business with Blackwell Publishing.

Registered office: John Wiley & Sons, Ltd, The Atrium, Southern Gate, Chichester, West Sussex, PO19 8SQ, UK

Editorial offices: 9600 Garsington Road, Oxford, OX4 2DQ, UK
The Atrium, Southern Gate, Chichester, West Sussex, PO19 8SQ, UK
111 River Street, Hoboken, NJ 07030-5774, USA

For details of our global editorial offices, for customer services and for information about how to apply for permission to reuse the copyright material in this book please see our website at www.wiley.com/wiley-blackwell

Library of Congress Cataloging-in-Publication Data

Kearnes, Matthew.
Critical risk research : practices, politics, and ethics / Matthew Kearnes, Francisco Klauser, and Stuart Lane.
p. cm.
Includes index.
ISBN 978-0-470-97487-2 (cloth)
1. Environmental engineering. 2. Risk management. 3. Technology–Moral and ethical aspects. I. Klauser, Francisco Reto. II. Lane, Stuart N. III. Title.
TA170.K43 2012
361.1–dc23

2011047553

A catalogue record for this book is available from the British Library.

Wiley also publishes its books in a variety of electronic formats. Some content that appears in print may not be available in electronic books.

Set in 9.5/13pt Meridien by Aptara Inc., New Delhi, India
Printed and bound in Singapore by Markono Print Media Pte Ltd

First Impression 2012

Contents

Contributors

Louise J. Bracken is Reader in Physical Geographer at Durham University. She specialises in the science of fluvial geomorphology and the practices of interdisciplinary working through translating science into practical solutions to real-world environmental problems. Louise's research explores the complex relationships within rivers between the processes that generate and supply runoff and fine sediment, processes that move the water and sediment through the river system and how knowledge's of rivers are used in practice to manage the natural environment. This research matters to the management of river systems today and in the future, since under predicted climate change water/sediment systems will change significantly. Louise's work is shaping policy regarding river habitats, directs institutional ways of working with rivers and is establishing new research agendas. She is one of a few international physical geographers recognised for pioneering work in fluvial processes and simultaneously conducting leading interdisciplinary research across the natural and social sciences. She can be contacted at L.J.Bracken@durham.ac.uk.

Jason Chilvers is Lecturer at the Science, Society and Sustainability group, School of Environmental Sciences, University of East Anglia. His research focuses on relations between environment, science, policy and society, and spans studies of governance, appraisal, public understanding, and public participation in relation to science, technology and environmental risk issues. He has published widely on these themes in books, policy reports, and peer-reviewed journals such as *Science, Technology, and Human Values, Environment and Planning A* and the *Journal of Risk Research*. Recent publications include an edited special issue on networks at the science-policy interface (*Geoforum*, 2009) and *Sustainable Participation?* (Sciencewise, 2010). He is the director of an international ESRC seminar series on 'Critical public engagement' and recently served on the Royal Society Kohn Award Panel for Excellence in Engaging the Public with Science. He can be contacted at: Jason.Chilvers@uea.ac.uk.

Brian R. Cook is a postdoctoral researcher at the UNESCO Centre for Water Law, Policy and Science at the University of Dundee. His research explores the dynamic interactions between knowledge and behaviour in relation to the water/risk interface. He is interested in the power embedded in flood management knowledges and practices. This work has taken him from analyses of flood impacts in Canada, to national scale decision making in Bangladesh, to the non-governmental organisations that have come to mediate catchment management in the UK. He has co-edited a special issue of *Environmental Hazards*, which assembled key thinkers and practitioners to challenge prevailing narratives and assumptions of flooding and flood management in Bangladesh. He can be contacted at: b.r.cook@dundee.ac.uk.

Sarah R. Davies is a researcher at Arizona State University's Center for Nanotechnology in Society (CNS-ASU). Her research interests are in public engagement and public understandings of science, science in museums, and the governance of new and emerging technologies. Her PhD was carried out in Imperial College London's Science Communication Group; since then, she has worked at Durham University (in its Institute of Hazard, Risk and Resilience) and as a Public Engagement Fellow at Beacon North East before moving to CNS-ASU. She has published in journals such as *Science Communication, Science as Culture* and *Public Understanding of Science*, and has co-edited three volumes *Science and Its Publics* (Cambridge Scholars Publishing, 2008),

Understanding Public Debate On Nanotechnologies: Options For Framing Public Policies (European Commission, 2010) and *Understanding Nanoscience and Emerging Technologies* (Akademische Verlagsgesellschaft, 2010). She can be contacted at: Sarah.Davies@asu.edu.

Lena Dominelli holds a Chair in Applied Social Sciences in the School of Applied Social Sciences and is Associate Director at the Institute of Hazards, Risk and Resilience Research at Durham University where she Heads the Programme on Vulnerability and Resilience. She currently holds (as PI) a Major ESRC funded project entitled 'Internationalising Institutional and Professional Practices' and (as CI) another significant EPSRC funded project entitled, 'Climate Change, the Built Infrastructure and Health and Social Care Provisions for Older People'. Alongside the wealth of experience she has had as a university educator and researcher, she has worked in social services, probation and community development. She has published widely in social work, social policy and sociology. Several of these are classics and have been translated into many languages. She is recognised as a leading figure in social work education globally. Her latest book is entitled *Green Social Work*. Professor Dominelli was elected President of the International Association of Schools of Social Work (IASSW) from 1996 to 2004, and is currently chairing the IASSW Committees on Disaster Interventions and Climate Change and is representing the social work profession at the United Nations discussions on climate change, including those to be held in Durban, South Africa from 29 November to 12 December 2011. She has also been the recipient of various honours including a Medal in 2002 for her contribution to social work given by the Social Affairs Committee of the French Senate and an honorary doctorate in 2008 from the University of KwaZulu-Natal in Durban, South Africa. She may be contacted at: lena.dominelli@durham.ac.uk.

Carl Grundy-Warr is a Senior Lecturer at the Department of Geography, National University of Singapore. He has engaged in long-term fieldwork within mainland Southeast Asia on political geographies of forced displacement, borderlands, environmental resource politics, and the human geographies of natural hazards. He has published in numerous journals and co-edited *Borderscapes: Hidden Geographies* and *Politics at Territory's Edge* with Prem Kumar Rajaram (University of Minnesota Press, 2007). He is currently engaged in international collaborative projects on livelihood and environmental security in the Mekong Basin, and geographies of public health in wetlands associated with food-borne parasites. He may be contacted at geocerg@nus.edu.sg.

Benjamin Horton is an Associate Professor in the Department of Earth and Environmental Science at the University of Pennsylvania. He works in the Sea Level Research Laboratory and focuses on the relationships between climate and sea level change seeking to better understand the external (such as sea-level and climate change, earthquakes and tsunamis) and internal mechanisms (including the coastal sedimentary budget) that contribute to the sea-level changes we observe and reconstruct. He may be contacted at: bphorton@sas.upenn.edu.

Matthew B. Kearnes is a Senior Lecturer in Interdisciplinary Environmental Studies in the School of History and Philosophy at the University of New South Wales, Australia. His research focuses on understanding the role of scientific and technological thinking in the politics and policies of contemporary democracies. Focusing particularly on research in areas such as nanotechnology and synthetic biology, he has authored, with Phil Macnaghten and James Wilsdon, *Governing at the Nanoscale: People, Policies and Emerging Technologies* (Demos, 2006) and co-edited a special issue of *Science as Culture* on *(Re)Imagining Nanotechnology.* He can be contacted at: m.kearnes@unsw.edu.au.

Francisco R. Klauser is Assistant Professor in political geography at the University of Neuchâtel, Switzerland. His work focuses on the relationships between space, surveillance/risk and power, with a particular focus on public urban space and places of mobility. His research interests also include urban studies and socio-spatial theory. In recent years, Francisco Klauser has developed an international portfolio of work on issues of security and surveillance at sport mega-events and in the aviation sector. He can be contacted at: francisco.klauser@unine.ch.

Stuart N. Lane is Professor of Geomorphology at the Institut de Géographie, Faculté des Géosciences et l'Environnement at the Université de Lausanne, Switzerland. He is a geographer with some training in civil engineering and has won many prizes for his research, concerned with modelling and remote sensing of river flow, sediment and solute transport, river ecology and floods. He is the Editor-in-Chief of the journal *Earth Surface Processes and Landforms* and has co-edited a number of recent publications including, *Landform Monitoring, Modelling and Analysis* (1998, Wiley), *High Resolution Flow Modelling in Geomorphology and Hydrology.* (1999, Wiley Advances in Hydrology Series) and *Computational Fluid Dynamics: Applications in Environmental Hydraulics* (2005, Wiley). Stuart is particularly interested in the ways in which the practice of science can be democratised particularly in the field of risk management, and is currently engaged in a number of research projects in this area. He can be contacted at: stuart.lane@unil.ch.

Lisa Law is Senior Lecturer in the School of Earth and Environmental Sciences, James Cook University, Cairns. She is a cultural geographer with interests in the relation between people, place and identity – mostly in Southeast Asia – and has taught in Australia, Singapore and the United Kingdom. She is author of *Sex Work in Southeast Asia: The Place of Desire in a Time of AIDS* (Routledge, 2000), and co-editor, with Ien Ang and Mandy Thomas, of *Alter/Asians: Asian-Australian Identities in Art, Media and Popular Culture* (Pluto Press, 2001) and, with Lily Kong, a special issue of *Urban Studies* titled "Contested landscapes, Asian cities" (2002). She has recently taken up a post as an Editor of *Asia Pacific Viewpoint*, a Wiley-Blackwell journal publishing articles in geography and allied disciplines about the Asia Pacific region. She can be contacted at lisa.law@jcu.edu.au.

Phil Macnaghten is Professor of Geography at Durham University. His research focuses on the ethical and societal dimensions of new science and technology, public deliberation and anticipatory governance, narrative approaches to policymaking, and the study of socio-nature. His principal publications include: *Contested Natures* (Sage, 1998), *Bodies of Nature* (Sage, 2001), *Governing at the Nanoscale* (Demos, 2006), *Reconfiguring Responsibility* (European Commission, 2009) as well as a number of edited collections and papers. He contributes to debates on the governance of science and technology in the UK, Europe and Brazil and is a member of the EPSRC's Strategic Advisory Network. He can be contacted at: p.m.macnaghten@durham.ac.uk.

Claudia Merli is Lecturer in Anthropology at Durham University. She specialises in medical anthropology. Her research on disasters focuses on the religious and theological discourses following natural hazards, and the increasing application of PTSD diagnostic category across to address disasters' consequences in post-disaster local populations. She conducted field research on Islamic and Buddhist theodicies in Southern Thailand in the aftermath of the 2004 Indian Ocean tsunami (December 2004–March 2005; March–June 2006). She is author of *Bodily Practices and Medical Identities in Southern Thailand* (Uppsala University Press, 2008). She can be contacted at claudia.merli@durham.ac.uk.

Katie J. Oven is a post-doctoral research associate in the Department of Geography and the Institute of Hazard, Risk and Resilience (IHRR), undertaking applied, interdisciplinary research on 'natural' hazards and development. Katie has field experience in Nepal, Taiwan, New Zealand and more recently the UK. Her doctoral research investigated the vulnerability and resilience of rural communities to landslides in the Nepal Himalaya using a mixed methods approach. Katie's research interests include combining local and outside 'scientific' knowledge for disaster risk reduction (DRR), and the governance of risk and resilience at community and national scales. She has recently completed a NERC/ESRC-funded scoping study as part of the Increasing Resilience to Natural Hazards programme exploring these issues in the context of seismic-related hazards in Nepal. Katie is currently a PDRA on a multi-disciplinary EPSRC-funded project which aims to develop strategies to ensure the infrastructures and health and social care systems supporting the wellbeing of older people in the UK will be sufficiently resilient to withstand the impacts of extreme weather events under conditions of climate change. She can be contacted at k.j.oven@dur.ac.uk.

Jonathan Rigg is a Professor of Geography at Durham University. He has long field experience in Southeast Asia, working mainly on issues of rural development encompassing such themes as rural-urban relations and interactions, migration and mobility, and sustainable livelihoods. He has authored and edited a number of books including *Southeast Asian Development: Critical Concepts in the Social Sciences* (Routledge, 2008), *An Everyday Geography of the Global South* (Taylor and Francis, 2007), *Living with Transition in Laos: Market Integration in Southeast Asia* (Routledge, 2005), *Southeast Asia: The Human Landscape of Modernisation and Development* (Routledge, 2003), and *More than the Soil, Rural Change in Southeast Asia* (Prentice Hall, 2001). He can be contacted at: j.d.rigg@durham.ac.uk.

Jean Ruegg was trained as a geographer and an urban planner. He is now a professor of land use policies at the University of Lausanne. He is particularly interested in the dynamics operating between the production of territorial relations and the mechanisms designed for their regulation. He is the co-editor, with Simon Richoz and Louis-M Boulianne of *Santé et Développement Territorial: Enjeux et Opportunités* (PPUR, 2010). He can be contacted at jean.ruegg@unil.ch.

May Tan-Mullins is a lecturer with the International Studies division of The University of Nottingham, Ningbo, China campus. Her research interests are non-traditional security issues such as environmental, food and livelihood security matters. She recently co-edited, with Victor Savage, *The Naga Challenged: Southeast Asia in the Winds of Change* (Eastern Universities Press, 2005). She can be contacted at may.tan-mullins@nottingham.edu.cn.

Preface

This volume was born of a collaboration of a group of scholars connected to the Institute of Hazard, Risk and Resilience, Department of Geography, Durham University. The breadth of subject matter covered in this volume is a testament to the intellectual scope of the institute and extraordinary scholarly energy and enthusiasm it enables. The publication of this volume is also a timely reminder of the intellectual breadth of 'risk research'. Included in this collection are papers dealing with the aftermath of natural hazards, the risks of new technology and the increasingly interconnected strategies adopted to 'secure' public spaces. As the asymmetric threats of global terrorism and the novel risks of new technologies continue to occupy the contemporary political imagination, alongside the threats posed by natural hazards, the scope of this collection is indicative of the ways in which risk research has become a key site of interdisciplinary exchange between often diverse approaches and intellectual traditions. As editors we are therefore indebted to all of the contributors to this volume. It was only their enthusiasm and energy that has made this project possible. Thanks are also due to Rachael Ballard, Izzy Canning and Fiona Woods at Wiley-Blackwell for the work they have both devoted in making the publication of this volume possible. We are grateful to Cosette Stirnemann at Neuchâtel University for formatting the chapters of this collection.

Matthew Kearnes
Francisco Klauser
Stuart Lane
March 2012

CHAPTER 1

Introduction: Risk Research after Fukushima

Matthew B. Kearnes[1], *Francisco R. Klauser*[2] *& Stuart N. Lane*[3]

[1]School of History and Philosophy, University of New South Wales, Australia

[2]Institut de Géographie, Faculté des Lettres et Sciences Humaines, Université de Neuchâtel, Neuchâtel, Switzerland

[3]Institut de Géographie, Faculté des Geosciences et de l'Environnement, Université de Lausanne, Lausanne, Switzerland

This book had its origins when all three of us were closely connected with Durham University's Institute of Hazard, Risk and Resilience. The Institute was established through a combination of university and philanthropic funding, so as *'to make a difference to those who live with risk'*. This book reflects a shared sense that this moral imperative, that is common in contemporary risk research, and is generally considered to be benign, deserved a deeper and much more critical scrutiny. Indeed, we argue that risk research, as well as risk analysis and management more generally, fulfills an institutional role, tasked with reducing the loss of life, expressing a duty of care, enhancing health and well-being and increasing economic security. Such moral imperatives may be laudable, but they are equally bound to a set of other precepts and taken for granted assumptions: that risk are inately calculable and; that we need institutions with the necessary expertise to do these studies and calculations for us; that those institutions should communicate what they have found and calculated; that risks are determinate in the sense that they are knowable even if not known; that risk can be approached objectively, independent from other ways of knowing the world, such as through systems of belief; and ultimately that the analysis and management of risk exists for the greater good. This book is about looking at these precepts critically and throughout we advance a notion of 'critical' risk research.

Our presumption is not that risk research is inherently uncritical. Rather, we argue that the intellectual foundations of contemporary risk research need more critical attention. This raises a series of fundamental questions: What *are* risks and how do we relate to them? How are we framing,

---

Critical Risk Research: Practices, Politics and Ethics, First Edition.
Edited by Matthew Kearnes, Francisco Klauser and Stuart Lane.

approaching and studying risks, and what are the implications of these framings? What do we know and do about risks, and in the name of risks? This book critically addresses these questions. Yet in so doing, we do not attempt to offer a best-practice model of how risk research should be done. Rather, the book's ultimate objective is an attempt at self-reflective transgression. Through illustration, we aim to challenge the ways in which risk-problems are approached and presented, both conceptually by academics and through the, often implicit risk-framings that are encoded in the technologies and socio-political and institutional practices surrounding contemporary risk research and management.

Fukushima: lessons and challenges

In compiling this volume throughout 2010-2011 it has been impossible to avoid the catastrophic events being played out in North Eastern Japan where, on 11 March 2011, the world awoke to news of the Tōhoku earthquake, a magnitude 9.0 (Mw) undersea megathrust quake with an epicentre approximately 129km east of the Japanese city of Sendai.[1] Regarded as the most devastating earthquake recorded in Japan since the 1923 Great Kanto Earthquake, it generated tsunami waves with reported heights of 40m (Tekewaki 2011). Felt across the Pacific, the tsunami waves breached flood defences across a large area of the North East of Japan, flooding cities and destroying infrastructure. Much like the 2004 Indian Ocean Tsunami, international media coverage of the Tōhoku earthquake and tsunami was dominated by haunting images of flooded cities, devastated communities and twisted flood defences, together with reports of almost incomprehensible numbers of human causalities.

In the following days, and after a series of significant aftershocks,[2] it was revealed that the combined effects of the earthquake and tsunami had caused critical equipment failures and nuclear meltdowns at the Fukushima 1 Nuclear Power Plant, resulting in the release of radioactive material and frantic efforts to both contain the damage and to evacuate civilians from the region immediately surrounding the plant. While these events threatened to send Japan back into a financial depression,[3] the political fallout was felt internationally, with significant protests in Germany, Switzerland and Italy over the continuing reliance on civil nuclear power.[4]

---

[1] http://earthquake.usgs.gov/earthquakes/eqinthenews/2011/usc0001xgp/#details

[2] http://ihrrblog.org/2011/04/15/japan-still-shaken-by-aftershocks/

[3] http://edition.cnn.com/2011/BUSINESS/03/14/japan.quake.economy.monday/index.html

[4] http://www.dw-world.de/dw/article/0,,14939216,00.html

As these events played out during the completion of this volume, we reflect here on the important lessons we might draw for contemporary risk research about the nature of 'critique', before outlining the structure and plan of the volume.

Vulnerability of techno-scientific 'risk societies'

The most obvious lesson to be drawn from these events is that a quarter century after the tragedy at the Union Carbide pesticide plant in Bhopal, the core meltdown at Three Mile Island and the Chernobyl disaster, the Tōhoku earthquake and the meltdown at the Fukushima nuclear power station reveal the continuing potential for such incidents to fundamentally disrupt social, economic and political life. What is perhaps becoming progressively more extreme is the ease by which scenes of devastation can be geographically diffused such that the experience of those affected by such events is reproduced, albeit through very different and highly mediated means, in almost real time. Instantaneously, they bring the susceptibility of social infrastructures to catastrophic and devastating natural and technological hazards to the fore.

However, a generation after the emergence of critical interpretations of conventional risk analyses (Beck 1992; Brickman, *et al.* 1985; Douglas and Wildavsky 1982; Perrow 1984; Wynne 1996) the events in Sendai and Fukushima reveal much more than just our continuing vulnerability to these events. They also reveal our continuing dependence on conventional risk analyses, a set of failures that expose the assumptions upon which they are constructed and, above all, the paucity of our conceptual and practical tools for understanding, approaching and, eventually, living with the daunting existence and prospect of such events. Thus, in addition to providing an allegory of modern vulnerability, the Tōhoku earthquake and the meltdown at the Fukushima nuclear power station, reveal a significant set of analytical and empirical challenges for contemporary risk research, three of which shall be outlined below.

The nature and causes of risk

The first broad challenge arising from the Fukushima tragedy concerns our very understanding of the nature and causes of risks. More specifically, these events dramatically underscore the problems associated with the two (apparently trivial and often taken for granted) oppositions between 'normal' and 'exceptional' risks, and between 'natural hazards' and 'human agency'.

Though the events witnessed in Japan in March 2011 were, by any standard, extraordinary in their severity and magnitude they were not unexpected. Commenting shortly after the Tōhoku earthquake, Petley (2011) suggested that from a "geological perspective . . . these events were far from unusual taking into account the seismic history of the region". Indeed

Petley went on to suggest that "as far as I can see this earthquake, and the resultant tsunami, are remarkably unsurprising. They are exceptionally large for sure, and they were not predictable, but they are not beyond the bound of human experience in any way that I can see".[5] If the Tōhoku earthquake, though extreme in its magnitude, is consistent with the seismic history of the region, what of the resulting tsunami? In his study of the cultural memories of tsunamis in Japan, Smits (2011) notes "that large tsunamigenic earthquakes have occurred repeatedly in precisely the areas devastated by the March 11, 2011 event". Smits goes on to suggest that despite recorded incidents of events of similar scale and magnitude, and latent cultural memories and folklore, urban infrastructures in these regions were designed to withstand more frequent incidents of lower magnitude.

This fact points to the particular normalising effect of institutionalised risk research and practices of risk management. In the terms of classical risk analysis the devastation witnessed in Sendai and other Japanese cities is a reminder that "once again it is our preparedness that is at fault. Once again our knowledge of the hazard has failed to transfer into effective mitigation" (Petley 2011). The fact that these events are consistent with the seismic history of the region points to the enduring vulnerability of our existing social, political and economic infrastructures to low-frequency but high-impact events.

In his study of high-risk technologies Perrow (1984) notes that accidents and risks are a systematic – or 'normal' – feature of societies that are 'tightly coupled'. What he means by this is that societies where everyday interactions depend on largely invisible electrical power systems, telephone connections and data networks, are particularly susceptible to infrastructure faults that cause more systemic breakdowns.[6] He suggests:

> When we have interactive systems that are... tightly coupled, it is "normal" for them to have this kind of an accident, even though it is infrequent. It is normal not in the sense of being frequent or being expected-indeed, neither is true, which is why we were so baffled by what went wrong. It is normal in the sense that it is an inherent property of the system to occasionally experience this interaction.... We have such accidents because we have built an industrial society that has some parts, like industrial plants or military adventures, that have highly interactive and tightly coupled units. Unfortunately, some of these have high potential for catastrophic accidents. (p. 8)

---

[5] Petley makes this argument on the basis of an historical analysis of the seismicity of the region. See: Rhea, *et al.* (2010).

[6] Also see Graham, 2009.

Though incidents, such as the nuclear meltdown and release of radioactive material at the Fukushima Nuclear Power Plant, and the ensuing political and economic crises, are precipitated by a 'natural' disaster these events demonstrate how risk is equally, if not more acutely, produced by the coupling between system elements, including environmental hazards, management systems and technologies. In this analysis vulnerability is not simply attributable to any one element of the system, so precluding mechanistic analysis of causation. Rather, the system becomes vulnerable because of the connections between elements that may be hidden and dynamic, making them difficult to identify except with the benefit of hindsight. The risks and vulnerabilities induced by events such as the Tōhoku earthquake operate as a complex assemblage of social, political, technical and geological factors (Anderson, *et al.* 2012; Bennett 2005).

The coupling, and indeed inseparability, of these events also demonstrates the paucity of our conceptual vocabulary. As implied above, contemporary risk research has relied on a simplistic understanding of vulnerability – coupled with mechanistic notions of causation – which sees risks as originating in the inanimate, non-human world and whilst human action is conceptualised as exacerbating its effects and the vulnerability of human populations (Jasanoff 1999). This simplistic conception of the causes of risk and vulnerability is typically represented as some variation of the pseudo-formula: risk = hazard x exposure x vulnerability or risk = probability x consequence.[7] This formulation gives a veneer of technicality to a categorical distinction between 'natural hazards' and 'human agency'. If ever any more evidence is needed, what the events at the Fukushima power plant reveal is the conceptual redundancy of this dualism between 'natural hazards' and 'man-made risks'. The conceptual terminology that underpins this distinction – that risks and hazards can be distinguished on the basis of their primary 'origin' – has proved to be fundamentally ill-equipped to deal with the tightly-coupled vulnerabilities of social, political and technical infrastructures to catastrophic failures.

Socio-political ambivalences of risk

This conceptual failure also highlights a second set of challenges arising from these events for critical risk research. The Tōhoku earthquake and the meltdown at the Fukushima power station also reveals the ambivalent role that risk research itself – and particularly institutionalised forms of risk management and risk assessment that thrive upon this research – plays in producing these forms of social vulnerability. Though classically

[7] This formulation of social vulnerability to risk has been the subject of extensive critical commentary. See for example, Bankoff, *et al.* (2004).

understood as providing technical capacities for calculating risk probabilities and intensities, and predicting exposure pathways and patterns, the events in Japan expose the degree to which formal processes of risk analysis often form part of institutionalised cost-benefit calculations engaged in the construction of disaster preparedness infrastructures. Though the possibility – indeed the likelihood – of tsunami waves of similar levels were both a feature of local folklore, and predictable on the basis of the region's underlying seismology (Atwater, *et al.* 2005), the construction of flood defences and the positioning of nuclear power stations in Japan has been influenced by a range of additional social and political factors. Principle among these are local political debates about power plant siting (Hayden 1998; Juraku, *et al.* 2007) and the inevitable cost-benefit trade-offs involved in the construction of flood defences.

These events point to a broader lesson for risk researchers – as they reveal the degree to which institutionalised forms of risk analysis are often part of social and political systems that produce and intensify vulnerabilities to hazards and disasters. Risk assessments are given a preeminent role in formal planning processes and the associated political and economic calculations, often because it is presumed that such assessments are both unambiguous and unbiased. However, the analysis of risk assessment in practice reveals that it has to be highly constrained by both policy and institutions in order to make problems scientifically tractable and politically and socially manageable (Lane *et al.*, 2011). The critical danger for risk researchers is that, rather than mitigating the effects of these incidents, such research forms part of the institutional structures that force problems to become tractable in particular ways and, even, render social groups more susceptible to systemic harm.[8]

Scales of risk

The third critical challenge that the events surrounding the Tōhoku earthquake and the meltdown at Fukushima pose for contemporary risk research concerns the issue of scale. Assessments of the scale of disasters are fundamental to risk research, and more broadly are part of the ways in which societies make sense of troubling and disturbing events. In the immediate aftermath of the events in North Eastern Japan the initial response by international organisations and relief agencies was to produce maps. Maps of the earthquake zone, the frequency and magnitude of the aftershocks, the scope of tsunami inundation and the extent of radiation

---

[8] This argument is laid out in more depth in Lane, *et al.* (2011). The authors also develop an alternative and participatory model of risk research, which provides a response to these dynamics. See also, Lane (this volume).

release became the dominant way in which international observers made sense of these events and coordinated responses (see Figures 1.1, 1.2 and 1.3 for examples of the kind of maps produced in the days immediately after the disaster). As the international media struggled to communicate the sheer scale and complexity of the disaster they also resorted to comparisons with similar events and raw calculations of the expected numbers of human causalities and predicted economic losses. The event became represented as the biggest recorded earthquake in Japan, the most severe tsunami in living memory, the worst nuclear incident in Japan

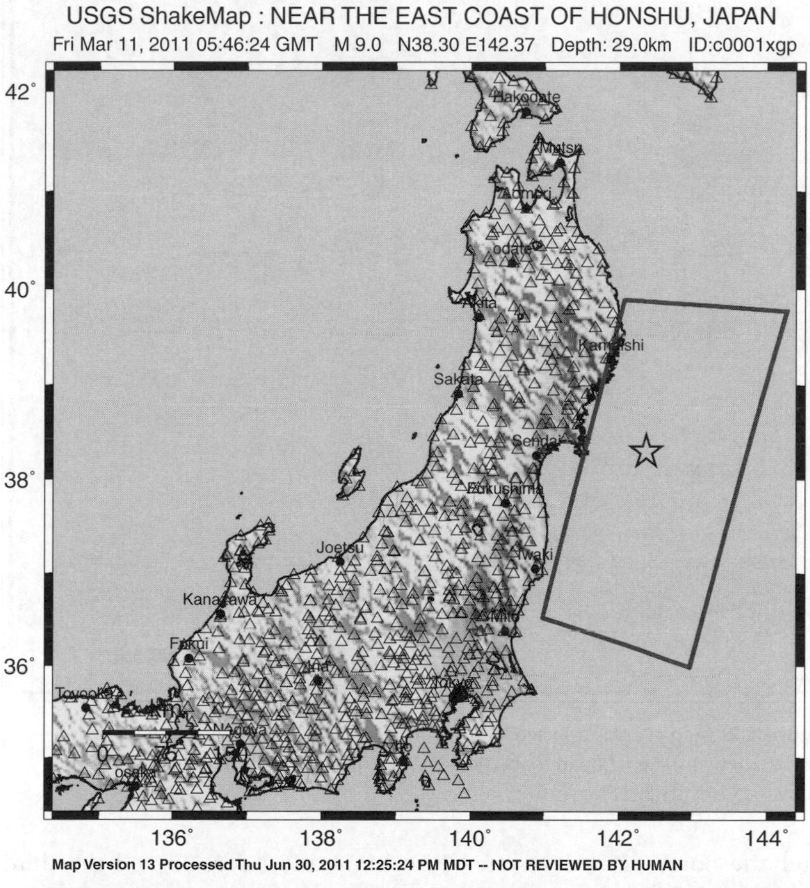

PERCEIVED SHAKING	Not felt	Weak	Light	Moderate	Strong	Very strong	Severe	Violent	Extreme
POTENTIAL DAMAGE	none	none	none	Very light	Light	Moderate	Moderate/Heavy	Heavy	Very Heavy
PEAK ACC.(%g)	<.17	.17-1.4	1.4-3.9	3.9-9.2	9.2-18	18-34	34-65	65-124	>124
PEAK VEL.(cm/s)	<0.1	0.1-1.1	1.1-3.4	3.4-8.1	8.1-16	16-31	31-60	60-116	>116
INSTRUMENTAL INTENSITY	I	II-III	IV	V	VI	VII	VIII	IX	X+

Figure 1.1 Shake Map of Tōhoku earthquake. *Source:* http://earthquake.usgs.gov/earthquakes/shakemap/global/shake/c0001xgp (accessed, 13 July 2011).

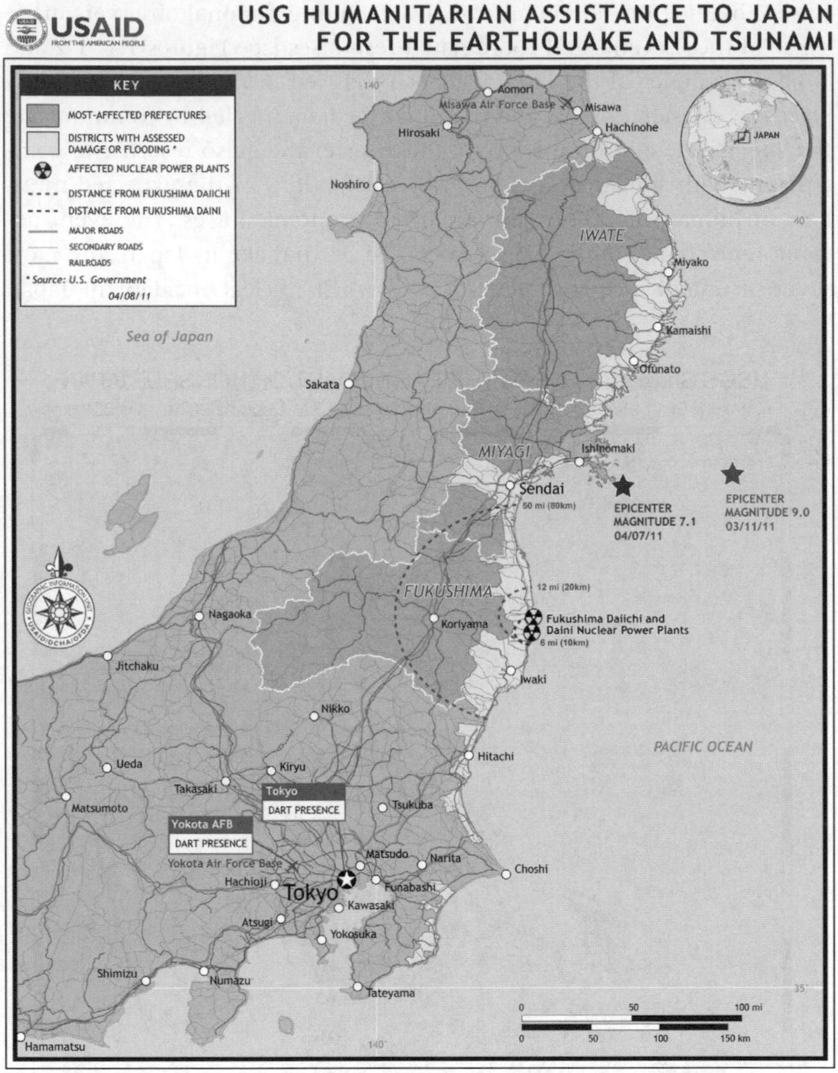

Figure 1.2 Map of areas affected by of Tōhoku earthquake and ensuing tsunami. *Source:* www.usaid.gov/japanquake/04.08.11-JapanEarthquakeTsunamiMap.pdf. (accessed, 13 July 2011).

after the bombs that devastated Hiroshima and Nagasaki and the third most significant nuclear accident after, Chernobyl and Three Mile Island. However, what was lost in this reporting, particularly as the events in Japan were translated into implications for other nations and regions, was their historical and geographical specificity. 'Explanation', through this kind of reporting, stripped away the deeper, more integrated and complex set of factors that had precipitated the tragedy, and in so doing further perpetuated simplistic and naïve assessments of what should have been done differently.

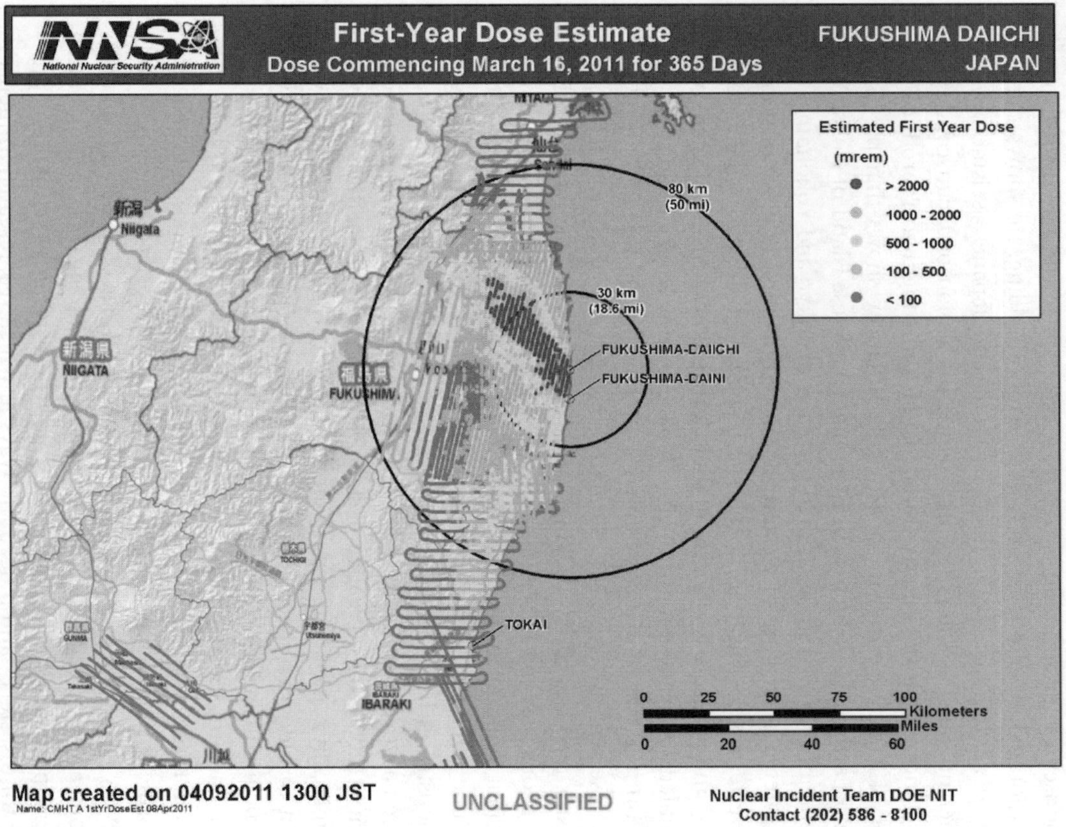

Figure 1.3 Estimated Radiation Dose Map. *Source:* http://blog.energy.gov/content/situation-japan (accessed, 13 July 2011).

In comparison to these initial attempts to attend to the scale of the Fukushima disaster, what these events reveal is the importance of a range of contextual factors that shape existing disaster preparations and emergency responses and thereby contribute to the systemic social and infrastructural vulnerabilities. For example, in his study of the 'cultural politics of Japanese seismicity' Clancey (2006) demonstrates that historic responses to the seismology of Japan have, since the mid 1800s, been incorporated in nationalist projects of nation-building. He shows the degree to which Japanese architectural style and Japanese seismology together operated as filters through which the nation of Japan conducted its relations with the international world and sought to reproduce notions of Japanese nativism. In particular, Clancey shows that the Great Kantō Earthquake, which struck the main Japanese island of Honshū on 1 September 1923 and devastated the cities of Tokyo and Yokohama, was interpreted as a failure of western-designed buildings. In place of notions of western modernisation, a nativist response to the earthquake dominated and particularly the notion that "foreign knowledge had again been humbled by Japanese nature, while Japanese knowledge . . . had again ridden out the waves" which resulted in a "narrative of foreign failure and Japanese tenacity" (p. 223). Similarly, in his study of more recent post-disaster recovery efforts, Edgington (2010) also finds a similar set of intersections between earthquake recovery programmes and contemporary national political imperatives. In a detailed account of the rebuilding of Kobe after the Great Hanshin-Awaji Earthquake in 1995, Edgington demonstrates how the destruction caused by the earthquake was transformed into an opportunity for political and economic renewal, what Rozario (2007) calls 'creative destruction'. Though these efforts were geographically varied, Edgington shows that "in Kobe, opportunities were taken to redevelop the older, inner parts of the city. Opportunities were also taken during the recovery to build new economic infrastructure so as to gain a comparative advantage over other cities in Japan" (p. 14).

What these two studies point to is the way that the science of formal risk assessment, and the expert judgements built into disaster preparedness and recovery initiatives, are typically a product of a range of social and political factors, actively reflecting and involved in reproducing social and political orders, particularly at the national scale. The ways in which nations prepare for disasters – and indeed the ways in which risk research influences urban planning policy – operates as a critical site for national myth making and political reproduction. This intersection between the technical and the political is made all the more obvious in responses to the nuclear meltdown at the Fukushima nuclear power plant, particularly given the persistent accusations of secrecy levelled at both the Japanese government and the Tokyo Electric Power Company (TEPCO) (Onishi and Fackler 2011).

In the context of existing public concerns about nuclear power in Japan, official responses to the nuclear meltdown are an indication of the political stakes at play. For example, Nelson (2011) suggests that "the bombings of Hiroshima and Nagasaki, fallout from the testing of Soviet nuclear weapons, and the Lucky Dragon Incident of 1954 left the Japanese in the 1950s with what some observers have called a 'nuclear allergy'. Historically, Japanese anti-nuclear-weapons activists have been among the most vigorous in the world". In spite of these incidents during the 1950s a political decision to engage in the development of civil nuclear power programme was taken, driven by the desire for energy security and the attempt to 'civilise' nuclear energy through the 'Atoms for Peace' programme. In light of broad public opposition, and intense local political debates about the siting of nuclear power stations (Hayden 1998), the Japanese government launched a programme of public relations initiatives to convince the population of the merits of nuclear power.

Though the recent events at the Fukushima plant are likely to spark off political debates about the safety of civil nuclear power, allegations of official secrecy demonstrate the ways in which response mechanisms are shaped by long-term political imperatives. Though this may be expected in any area of critical national infrastructure or strategic political priority, what is significant about this for risk researchers is the ways in which formalised risk assessment is invoked as part of an intensely political process. Take, for example, the conflicting advice offered on the declaration of an exclusion zone around the Fukushima reactor in the days immediately following the meltdown. Whilst world nations advocated an exclusion zone of 80kms for their nationals, the Japanese authorities initially advised on a 20km radius zone – a strategy designed to limit both the atomic and political fallout of the Fukushima meltdown. Of course we are not debating the merits of each of these strategies, but rather pointing to ways in which this example is indicative of the political co-constitution of contemporary risk research. In this case, official assessments of acceptable tolerance limits, and the predicted spread of radiation seem to have been influenced by the fraught state of Japanese environmental politics, and a desire to limit the political damage of the meltdown.

Critical risk research

The intersections between risk assessment, risk analysis and contemporary political power are an expression of what Jasanoff (2004) terms the co-production of science and social order. For Jasanoff, the conceptual terminology of co-production helps to clarify our understanding of both the 'social construction' of officially sanctioned forms of knowledge making and the constitutive role that such knowledges play in sustaining contemporary political cultures and social order. Jasanoff defines the term co-production in the following ways:

> Briefly stated, co-production is shorthand for the proposition that the ways in which we know and represent the world (both nature and society) are inseparable from the ways in which we choose to live in it. Knowledge and its material embodiments are at once products of social work and constitutive of forms of social life; society cannot function without knowledge any more than knowledge can exist without appropriate social supports. (p. 2)

In this light, what the incidents in Japan associated with the Tōhoku earthquake and the meltdown at the Fukushima power station amply demonstrate is the mutually co-constituting relationship between formalised risk assessment, disaster planning and contemporary political order. The strategies that political authorities adopt to both prepare for and recover from disasters are, in part, influenced by their historical and geographical specificity. They are an expression of a range of more-than-technical considerations. Jasanoff's terminology also suggests that the conduct of official expertise also confers a form of technical legitimacy on contemporary decision-making, that function to preserve the established topologies of political power often in light of critique and contestation. From this perspective, risk assessment operates as a set of institutionally sanctioned rituals deployed in maintaining state power and hegemonic interests (see also O'Malley 2004; Power 1997; Wynne 1982).

It is for this reason that we suggest that the co-production of risk management and contemporary political order poses a significant set of methodological, ethical and conceptual challenges for risk research, and indeed for risk researchers. This volume might therefore be read as a response to this challenge that attends to the pragmatics of risk research in both its messy complexity and often compromising institutional settings. We start from the premise that the nature of contemporary risks is that they are highly complex. Explaining the emergence, modalities and implications of different types of risks, and using this to inform effective and responsible intervention, requires appreciation of a multitude of possible contributory causes, assembled around the problems at hand. Necessarily, risk research interacts with various psychological, social, economic, institutional and political factors, whose role in defining and shaping the problems at hand is crucial (Short 1984).

However, even where a problem is approached from a range of perspectives, this is rarely sufficient to open up the ways through which risk problems are framed, studied, managed and 'solved'. Such framings are not subject to critical interrogation and scrutiny. It is for this reason that the papers that comprise this volume also make a second move, together arguing for a form of self-critical and reflexive risk research. If risk research is undertaken to address real-world problems, particularly as they are defined by people affected by disasters, the papers that comprise this volume argue that it is also critical to explore the often implicit ways in which risk research frames problems. Across a range of research methodologies and sites of empirical investigation, this volume makes the case for a critical turn in contemporary risk research, an approach to risk scholarship that attends to the social, political and economic contexts that shape the constitution of the field. The goal here is to render risk research itself as a site of critical enquiry.

Before outlining the structure of the book, we sketch three cross-cutting themes and detail how the papers that comprise this volume may be read as offering a set of insights of a critical risk research.

1. *Conceptualisation*

Modern risk research is in many ways a product of its time. The emergence of the concept of risk is associated particularly with the development of both system engineering and the insurance industry, including the formation of legal doctrines of accountability and compensation for harms (Ewald 1999; O'Malley 2004). Risk is, in this sense, inseparable from the notions of calculation and quantification, as strategies designed to manage and regularise unpredictable events necessarily entail an attempt to predict the scale and scope of potential threats. Our contemporary understanding of risk is also inseparable from the kinds of threats that are faced by modern societies – from global terrorism, systemic and catastrophic infrastructure failures to the latent potential for mundane technologies to herald unanticipated consequences. Ulrich Beck's theory of the emergence of a world risk society, which has significantly shaped conceptual treatments of risk and vulnerability, is also a theory of European modernity. He famously argued that events such as the Chernobyl disaster, the tragedy at the Union Carbide pesticide plant in Bhopal and the core meltdown at Three Mile Island have the capacity to induce a range of anthropological shocks (Beck 1987). These catastrophic events have the potential to challenge the ways in which risk is understood, compartmentalised and managed. Commenting on Chernobyl he suggests that this single event catalysed a new set of cultural meanings. For Beck, Chernobyl created the conditions for a broad realisation that the risks of nuclear power, itself deeply symbolic of the biopolar world of the Cold War, could not be contained by political, regulatory or geographical boundaries. The risks of nuclear power are literally carried on the wind. Secondly, Beck suggests that incidents like Chernobyl called into question the inseparability of risk from technological modernity. Risks were not simply the products of faults or accident but, as Perrow (1984) also argues, the systematic product of 'normal' conditions. Since Beck's original formulation a generation of risk scholars have demonstrated similar kind of dynamics at play in contemporary techno-environmental controversies around acid rain, BSE and genetically modified food (Lupton 1999).

In the aftermath of the Tōhoku earthquake and the meltdown and release of radiation at Fukushima it is clear, however, that Beck's notion of anthropological shock does not have the same traction. Though these events are deeply shocking, it is also evident that we have been here before. In the rush to provide some analysis of the scale of these events, by comparing them to Chernobyl and Three Mile Island, what was striking about the international media coverage was the degree to which the events in Japan fell into a range of now accepted disaster narratives (Erikson, 1995).

This is not to diminish the significance of these events, but rather to suggest that what the Tōhoku earthquake and Fukushima meltdown reveal is the poverty of current methodological and conceptual tools in risk research. The question is how to practice risk research 'after Fukushima' with a set of theoretical tools sensitive enough to the specificity of the events in question. The solution that we offer in this volume is practical and epistemological rather than philosophical. Rather than declare a new epoch, of a 'post-risk society' or the emergence of forms of collective neurosis for example (Isin 2004), in this volume we move toward a notion of a 'critical risk research'. Our ambition is to move beyond what we usually do; to look critically at the research that we are ourselves doing (Bourdieu, 2004), to raise awareness of the interests, agendas, impacts, ethical issues and power games producing, and produced by, risk research.

Two models of this kind of (self) critical risk research are evident in the papers that comprise this volume. The first is offered by Lane (this volume), who argues that in light of the co-production of research on flood risks and on the political and economic vulnerabilities induced by such events – what he terms an 'unethical trend' in risk research – researchers need to attend to a 'moral imagination' in their work. Lane offers a model of ethical risk research that seeks to avoid conceptualising the public as in need of education, what Wynne (1992) characterises as the 'deficit model' of public understanding of science. Part of a broader project, aimed at a radical re-thinking of risk research as a form of participatory knowledge creation (see for example, Lane, *et al.* 2011), Lane's provocation is that responding to the co-production of risk knowledge and political power should be viewed as an ethical responsibility for risk researchers. The second approach is outlined by Macnaghten and Chilvers (this volume) who explore current strategies aimed at generating democratic and participatory forms of risk governance that are being taken up in the governance and regulation of new technologies. In their chapter,

Macnaghten and Chilvers review the formation of the Sciencewise Expert Resource Centre, a UK government institution tasked with coordinating early and anticipatory forms of public engagement and participation on themes as diverse as stem cell research, nanotechnology and geoengineering. Whilst informed by a range of critical interventions in the value of upstream public engagement in shaping regulatory responses to new technologies (see for example Wilsdon and Willis [2004]), in practice a range of models of public dialogue are current. Macnaghten and Chilvers' contribution to the development of critical risk research is, like Lane, to argue that research practice should start with public values, designing interventions and regulatory structures that function to 'open up' rather than close down decision making processes to a range of societal voices (Stirling 2008).

2. *Disciplinarity and Interdisciplinarity*

In recent years, risk research has been characterised by two countervailing shifts. On the one hand the complexity of contemporary risk issues is increasingly understood as requiring both interdisciplinary and participatory research practices. In concert with a range of approaches that seek to develop a synthesis between the social and physical sciences and socially robust forms of knowledge making (Gibbons, *et al.* 1994), the value of diverse perspectives is also increasingly recognised in contemporary risk research. However as Bracken (this volume) and Rigg *et al.* (this volume) report, in the practice of risk research divisions between disciplinary approaches are often both maintained and reinforced. Both Bracken and Rigg *et al.* go onto to suggest that the persistence of these distinctions is in part explained by a continuing and often implicit dualism between the 'hard facts' of disasters – which are articulated technically – and the 'softer' analyses of the 'social aspects' and 'lay perceptions' of these events.

Given these contradictory shifts, the Tōhoku earthquake and the meltdown at the Fukushima power station amply demonstrate the paucity of these conceptual terms – and the continuing distinctions between the technical 'facts' of risk and the social and cultural 'values' invoked in situated sense making – for both understanding and responding to complex and multifaceted events. The key issue here is not simply overcoming a set of disciplinary distinctions by developing synthetic research practices. Rather the challenges posed by these events concern the adequacy of the basic conceptual terminology of risk research – risk, hazard, vulnerability, exposure and tolerance – and its very framings themselves. In practice these terms, though commonplace in risk research, are typically conceptualised very differently. For example, Davies *et al.* (this volume) outline the incompatibility of existing conceptualisations of these basic terms, where current approaches to risk research are inspired by contrasting interpretations of the foundational concepts of the field. As a consequence research objectives, goals and methodologies tend to be framed in contrasting terms.

In practice, to overcome these conceptual and methodological impasses requires much work and persistence. In this sense, the conceptual challenge posed by events such as the Tōhoku earthquake are not simply to develop a definitive conceptual terminology, but to proliferate a new set of concepts that enable such events to be represented in all their hybrid complexity, as neither simply 'natural hazards' or 'man-made risks' (Whatmore 2002).

The papers of this volume offer a set of resources for engaging in the hybrid disciplinary complexity of contemporary risks and hazards. For example, Merli (this volume), describes the wide-ranging anthropological literature that shows that disasters are more than just physical problems. Disruptive catastrophic events reveal the entwined relationship between nature and culture and so implicate cultural questions in their understanding. Dominelli (this volume) describes the role of social policy in addressing the much broader set of consequences arising from disaster, ones that go well beyond, those that can be counted or calculated.

3. *Institutionalisation*

As we introduced earlier, a key feature of contemporary risk research is its increasingly institutionalised role in the formal responses to disasters. In some contexts a disaster or risk 'industry' has arisen which trades on the kinds of unique skills and expertise that risk researchers are able to offer in developing disaster preparedness procedures and in post-disaster relief and reconstruction efforts. Rigg *et al.* (this volume) and Dominelli (this volume) are both parodies of this point. The 'urgency funding' willing to be provided by research councils, such as that obtained by Rigg *et al.* (this volume) to study the aftermath of the 2004 Asian tsunami in Thailand, is a good example. Taking a critical view of such initiatives is not the same as arguing that they should not happen. Rigg *et al.* (this volume) show how studying this event, aside from raising a set of deeper challenges for risk analysis and management, also reveals the critical problem of the risk 'academy' and how its disciplinary structure prevents the realisation of truly interdisciplinary approaches to risk management. We explore this issue further below but we emphasise that it is not that risks, and their manifestations as events, are an illegitimate focus of academic enquiry. Rather, we cannot be blind to the ethical questions that underpin them and the ethical consequences that arise from them, anticipated or not. If risk research is justified from a moral imperative, the trajectory of that research must be followed with a close and sensitive attention to the ethical difficulties that the research goes on to pose. Risk research is largely unregulated despite the fact that it can have a profound impact upon both risk and wider society (Macnaghten and Chilvers, this volume).

The point here is that the institutionalisation of risk research in these contexts has potentially unanticipated consequences for different kinds of research methods and approaches utilised and the broader political implications of knowledges produced through these kinds of risk research. Both Kearnes (this volume) and Klauser and Ruegg (this volume) make this point, suggesting that the processes whereby risk research is incorporated into formal processes of risk management have important social and political effects. Examining risk management strategies deployed at Geneva Airport, Klauser and Ruegg (this volume) explore the synchronicity between contemporary political discourses of securitization and the situated conduct of airport security and surveillance services. In addition they demonstrate the ways in which risk research – particularly in the form of academic expertise on security and criminality – form part of the socio-political processes and practices of managing risk. Kearnes (this volume) makes a similar argument, analysing the ways in which governance and regulation of technological risks, in this case the risks associated with nano-materials, is influenced by contemporary political rationalities. In this particular area of risk governance, Kearnes describes the ways in which anticipatory and pre-emptive strategies, originally developed as a response to the asymmetric threats of global terrorism, are beginning to influence approaches deployed in managing the latent and potential threats posed by novel materials. Both of these cases describe situations where the modal logics and techniques of risk management developed in one field, are extended beyond their original purpose. Viewed in co-productionist terms, this epistemological extension is indicative of the ways in which contemporary risk research forms part of the discursive structures that support and sustain contemporary political rationalities and hegemony. The ways in which risk research is increasingly institutionalised as part of a 'risk industry' begs a range of significant ethical questions for risk researchers – concerning the ways in which such knowledge contributes to institutionally sanctioned judgements that may actually function to increase rather than decrease social, political and economic vulnerabilities.

Structure of the book

Developing these themes across a range of contributions, this volume offers a collection of essays about what it means to do risk research, and about how – and with what effects – risk research is practiced, articulated and exploited. Following this broad objective, the book is divided into three core sections, with equal numbers of chapters, which together provide a focused discussion of the Practices (1), Politics (2) and Ethics (3) in risk research.

Part I of the book is entitled 'Practices in Risk Research'. It aims at the assessment and investigation of some of the main methodological issues in doing risk research. Comprised chapters aim to critically examine the meaning and practical implications of the concept of 'risk' itself, to problematise the logics and driving forces of our methodological approaches, with a particular focus on issues of interdisciplinarity, and to reflect upon the practicalities and pragmatics of risk research more generally. Broadly speaking, two main questions are at the core of these investigations: How are problems of risk shaping our methodological and conceptual approaches? And, in turn, how are the concepts and methods we use conditioning the formats and contents of our problem framing and output?

Part II – 'Politics in Risk Research' – examines the institutions, agents and interests that are surrounding and shaping our perception, experiences and studies of risk. Risk, in this context, is understood as a hard-fought resource, i.e. a competitive market of academic research, political intervention and economic exploitation. This means: risk research is necessarily and inextricably implicated in a complex grid of relationships of power. It is situated within and in turn influencing the socio-political processes and practices of managing risk in the contemporary world (Bradbury, 1998). From various perspectives, this part, hence, raises a series of important questions relating to the socio-political constructions of risk and to the various logics and issues involved in current risk policies.

Part III – 'Ethics and Risk Research' – examines the various and complex ethical issues involved in, and produced through risk research. Our starting point is that risk research is by no means a neutral, value-free field of study (Fischhoff 1995; Renn 1998). Values are reflected in how risks are measured, framed, described, qualified, perceived, experienced, etc. Generating knowledge and practices about risk is always mediated by a series of norms, problems, intentions, institutions and agents, which directly and indirectly shape the form, direction and content of our output. At the same time, the field of risk research itself – by its practices and knowledge – actively participates in the co-production of 'risk' as a series of problems to manage and 'solve'.

In closing the volume we suggest that these issues are crucial to address today, particularly in the context of events such as the Fukushima earthquake and tsunami. Such events demonstrate the residual vulnerability of social and economic infrastructures and the often compromising position of formalised risk assessment and risk research. In making the case for a form of 'critical' risk research we suggest that this volume might therefore be read as a response to these challenges; a response that takes as its primary site of critical reflection the institutionalised role that risk research and risk researchers play in mediating responses to such events.

References

Anderson, B., Kearnes, M., McFarlane, C., and Swanton, D. (2012) On assemblages and geography. *Dialogues in Human Geography* in press.

Atwater, B.F., Satoko, M.-R., Kenji, S., Yoshinobu, T., Kazue, U. and Yamaguchi, D.K. (2005) *The Orphan Tsunami of 1700: Japanese Cluse to a Parent Earthquake in North America*, United States Geological Survey, Reston, VA.

Bankoff, G., Frerks, G. and Hilhorst, D. (eds) (2004) *Mapping Vulnerability: Disasters, Development and People*, Earthscan, London.

Beck, U. (1987) The anthropological shock: Chernobyl and the contours of the risk society. *Berkeley Journal of Sociology*, **32**, 153–165.

Beck, U. (1992) *Risk Society: Towards a New Modernity*, Sage, London.

Bennett, J. (2005) The agency of assemblages and the North American blackout. *Public Culture* **17**(3): 445–465.

Brickman, R., Jasanoff, S. and Ilgen, T. (1985) *Controlling Chemicals: The Politics of Regulation in Europe and the United States*, Cornell University Press, Ithaca, N.Y.

Clancey, G. (2006) *Earthquake Nation: The Cultural Politics of Japanese Seismicity, 1868–1930*, University of California Press, Berkley.

Douglas, M. and Wildavsky, A. (1982) *Risk and Culture: An Essay on the Selection of Technical and Environmental Dangers*, University of California Press, Berkeley.

Edgington, D.W. (2010) *Reconstructing Kobe: The Geography of Crisis and Opportunity*, University of British Columbia Press, Vancouver.

Erikson, K. T. 1995: *A New Species of Trouble: The Human Experience of Modern Disasters*: W.W. Norton and Co, New York.

Ewald, F. (1999) The return of the crafty genius: an outline of a philosophy of precaution. *Connecticut Insurance Law Journal*, **6**, 47–79.

Fischhoff, B. (1995) Risk perception and communication unplugged: twenty years of process. *Risk Analysis*, **15**(2), 137–145.

Gibbons, M., Limoges, C., Nowotny, H., Schwartzman, S., Scott, P. and Trow, M. (1994) *The New Production of Knowledge: The Dynamics of Science and Research in Contemporary Societies*, Sage, London.

Graham, S. (ed) (2009) *Disrupted Cities: When Infrastructure Fails*, London, Routledge.

Hayden, L.S. (1998) *NIMBY Politics in Japan: Energy Siting and the Management of Environmental Conflict*, Cornell University Press, New York.

Isin, I. (2004) The neurotic citizen. *Citizenship Studies*, **8**(3), 217–235.

Jasanoff, S. (1999) The Songlines of Risk. *Environmental Values*, **8**(2), 135–152.

Jasanoff, S. (2004) The idiom of co-production, in *States of Knowledge: The Co-production of Science and Social Order* (ed S. Jasanoff), Routledge, London, pp. 13–46.

Juraku, K., Suzuki, T. and Sakura, O. (2007) Social decision-making processes in local contexts: an STS case study on nuclear power plant siting in Japan. *East Asian Science, Technology and Society: An International Journal*, **1**(1), 53–75.

Lane, S.N., Odoni, N., Landstroem, C., Whatmore, S.J., Ward, N. and Bradley, S. (2011) Doing flood risk science differently: an experiment in radical scientific method. *Transactions of the Institute of British Geographers*, **36**(1), 15–36.

Lupton, D. (1999) *Risk*, Routledge, London.

Nelson, C.R. (2011) 'The energy of a bright tomorrow': the rise of nuclear power in Japan. *Origins*, **4**(9).

O'Malley, P. (2004) *Risk, Uncertainty and Government*, Glasshouse, London.

Onishi, N. and Fackler, M. (2011) In nuclear crisis, cripling mistrust. *The New York Times*, 12 June 2011.

Perrow, C. (1984) *Normal Accidents: Living with High-Risk Technologies*, Basic Books, New York.

Petley, D. (2011) From a geological perspective, what is surprising about the Sendai Earthquake? http://ihrrblog.org/2011/03/16/from-a-geological-perspective-what-is-surprising-about-the-sendai-earthquake/ (accessed 29 July 2011).

Power, M. (1997) *The Audit Society: Rituals of Verification*, Oxford University Press, Oxford.

Renn, O. (1998) Three decades of risk research: accomplishments and new challenges. *Journal of Risk Research*, **1**(1), 49–71.

Rhea, S., Tarr, A.C., Hayes, G., Villaseñor, A. and Benz, H.M. (2010) *Seismicity of the Earth 1900-2007, Japan and Vicinity*, 1 map sheet, scale 1:5,000,000, U.S. Geological Survey Open-File Report 2010-1083-D, Denver, CO.

Rozario, K. (2007) *The Culture of Calamity: Disaster and the Making of Modern America*, University of Chicago Press, Chicago.

Short, J.F. (1984) The social fabric of risk: toward the social transformation of risk analysis. *American Sociological Review*, **49**, 711–725.

Smits, G. (2011) Danger in the Lowground: Historical Context for the March 11, 2011 Tōhoku Earthquake and Tsunami. *Asia-Pacific Journal*, **9**(20), May 16, http://www.japanfocus.org/-Gregory-Smits/3531 (accessed 29 July 2011).

Stirling, A. (2008) 'Opening up' and 'closing down': power, participation and pluralism in the social appraisal of technology. *Science, Technology & Human Values*, **33**(2), 262–294.

Tekewaki, I. (2011) Preliminary report of the 2011 off the Pacific coast of Tohoku Earthquake. *Journal of Zhejiang University-SCIENCE A (Applied Physics & Engineering)*, **12**(5): 327–334.

Whatmore, S. (2002) *Hybrid Geographies: Natures Cultures Spaces*, Sage, London.

Wilsdon, J. and Willis, R. (2004) *See-Through Science: Why Public Engagement Needs to Move Upstream*, Demos, London.

Wynne, B. (1982) *Rationality and Ritual: The Windscale Inquiry and Nuclear Decisions in Britain*, British Society for the History of Science, Chalfont St Giles.

Wynne, B. (1992) Public uptake of science: a case for institutional reflexivity. *Public Understanding of Science*, **2**, 321–337.

Wynne, B. (1996) May the sheep safely graze? A reflexive view of the expert-lay knowledge divide, in *Risk, Environment and Modernity: Towards a New Ecology* (eds. S.M Lash, B Szerszynski and B. Wynne), Sage, London, pp. 44–83.

PART 1
Practices in Risk Research

CHAPTER 2
Practices of Doing Interdisciplinary Risk-Research: Communication, Framing and Reframing

Louise J. Bracken
Department of Geography, Durham University, Durham, UK

Introduction

Risk shapes the fundamental basis of how we live our lives and interact in society. Risk is all pervasive within our environment, at a variety of scales and severity, on a daily basis; some of these risks we can avoid, but many others we learn to live with or choose to take. The field of risk research has thus become a 'truly interdisciplinary concern involving virtually all the social and natural sciences, health interests and professional programs' (Montz and Tobin, 2011, p1). This is implicit in the many and varied interpretations of risk which exist in the literature. Definitions of risk tend to be probabilistic in nature by considering firstly, the probability of occurrence of a hazard that acts as a trigger of one or more undesirable events and/or secondly, the probability of a disaster which combines the probability of a hazard occurring with some consideration of the likely consequence (i.e. a social impact) (Smith, 1996; Stenchion, 1997; Downing *et al.*, 2001; Brooks, 2003; Brooks *et al.*, 2005). This has led to an increasing number of calls for interdisciplinary approaches to bring research from different disciplines together to overcome a tendency for disciplinary specialisation (Hansson, 1999; Medd and Chappells, 2007; Carolan, 2008). Central here is learning to work with the messiness of complex real world problems rather than trying to simplify them (Donaldson *et al.*, 2010).

It is well recognised that interdisciplinary practice is challenging because of: differences in epistemologies, knowledges and methods (Evans and Marvin, 2004); different ways of formulating research questions (Oughton

Critical Risk Research: Practices, Politics and Ethics, First Edition.
Edited by Matthew Kearnes, Francisco Klauser and Stuart Lane.

and Bracken, 2009); differences in the nature of communication (oral and written) (Bracken and Oughton, 2006; Huby and Adams, 2009); and differing attitudes within and between disciplines as to the importance of interdisciplinarity (Lowe and Phillipson, 2009). These challenges have been exacerbated by expert knowledge becoming segmented and bureaucratised in recent decades; and compounded by the fact that science and technology are often grounded in the philosophy of reductionism, which tries to simplify and prise apart complex, real world problems into manageable chunks of disciplinary specific research (see for example Brewer, 1999; Karlquist, 1999; Balsiger, 2004; Lawrence and Deprès, 2004). One key aspect associated with taking an interdisciplinary approach to research is that it tends to follow the problem such as the risk associated with a particular hazard. In this way, the risk develops the language and frame used to explore it. The appropriateness of interdisciplinary research is based on the premise that its associated collaboration and networking will produce innovative concepts and methods to answer complex research questions that are beyond the expertise of individual disciplines (Nissani, 1997; Bruce *et al.*, 2004; Donaldson *et al.*, 2010). In effect the issue framing is co-produced by more than one discipline. However achieving good interdisciplinary research is not necessarily straightforward.

My own research has a strong focus of working with systems that incorporate risky behaviour; for example flood production, channel change and the threat of extinction to endangered species (e.g. Bracken and Kirkby, 2005; Bracken *et al.*, 2008; Bolland *et al.*, 2010). I have also been involved in a number of interdisciplinary research projects combining different disciplines within and across both social and physical science. This has resulted in a series of published papers exploring the practice of doing interdisciplinary research; highly relevant to the practice of undertaking risk research (e.g. Bracken and Oughton, 2006; Oughton and Bracken, 2009). There are theoretical and conceptual aspects to interdisciplinarity, but in previously published work Liz Oughton and I have argued that if we are to address issues of interdisciplinarity fully, we also need to concern ourselves with practice; with the methods by which these concepts are employed. In this chapter I draw on my interdisciplinary research experience to consider some of the challenges of interdisciplinary working with a view to supporting those explicitly researching risk. As outlined above I assume that many aspects of researching risk implicitly require an understanding of interdisciplinary ways of working; combining knowledges and practices from a range of disciplines. First the chapter examines some of the different ways in which disciplines work. It then focuses on two key aspects of interdisciplinary practice; the significance and use of language in the development and implementation of research; and the role of issue framing. These are not new problems but become increasingly relevant as researchers from diverse backgrounds and trainings attempt to work together to address questions related to risk.

Doing interdisciplinary research

At the outset of this chapter it is important to reflect on why people choose to undertake interdisciplinary research. In interviews conducted with 12 academics leading funded interdisciplinary projects everyone was keen to undertake this type of research to *'give something back'* and/or to *'make a difference'* (Oughton and Bracken, 2009). Justifications for undertaking interdisciplinary research were implicitly grounded in the notion that interdisciplinary research, by including the needs of others, is somehow more valuable. These interviews demonstrated that the predisposition to undertake interdisciplinary research is realised through three different, although not mutually exclusive, routes. First, this was through collaboration with people from different disciplines, where researchers remained centred within their own area of expertise but were willing to engage with and to trust the expertise of others. Second, it was realised through reading adventurously and developing understandings that allowed researchers to work critically with others in different disciplines, but without more formal collaboration. Finally, it could be realised by undertaking training in a completely new area or field, with some of those interviewed taking high-level qualifications in multiple fields of expertise. These three routes were supported by specific practices, such as working with physical objects (e.g. models), visiting places to experience how interdisciplinary research is practiced, and serendipity. The material I cover in this chapter is more relevant to those undertaking interdisciplinary research in conjunction with others, the first of these routes.

Language is important in doing interdisciplinary research because it is a primary interface between people with different expertise and training. Distinctions between specialisms are striking in terms of their epistemologies: the ways in which they develop research questions and the methodologies chosen to explore those questions. Interdisciplinary research requires an understanding of the disciplines themselves as well as an understanding of how to connect disciplinary knowledge (Karlquist, 1999). A simplistic view suggests that physical scientists treat the topic of study as an object whereas to the social scientist the topic of study is the subject. As a consequence physical scientists generally use methods to monitor and evaluate the object, in a sense that separates them from that object. Social scientists typically have a richer set of relationships with their subject, which may range from methods that mimic the separation of researcher from researched (as in a questionnaire sample) through to methods which recognise their own role and effect on the research subject. This in turn leads to disciplines expressing themselves in different ways, which in turn results in variations in the style of oral and written communication and ways of relating research to published arguments and debates; all of which can present difficulties for reporting discussion and results between and across different areas of disciplinary research.

Differences in the scale of focus of research may vary between disciplines, which can make it difficult to relate different kinds of knowledge. Stereotypically, Dalgaard *et al.* (2003) argue that social science disciplines tend to work at more regional levels, whereas physical sciences tend to work at smaller scales. However, both physical and social sciences work on a range of spatial scales from the plot/individual to the global, and such ranges exist within disciplines as much as they do between them. Similarly they operate on varying temporal scales. In an interdisciplinary context, bringing together individuals with different definitions of the scale necessary for enquiry may lead to different definitions, and result in gaps in information flows, and consequent misunderstandings. Scale variations produce very different starting points from which to view the environment and hence can lead to very different problem framings and to highly diverse research strategies (Bracken and Oughton, 2006). In theory, being interdisciplinary should circumvent this, if it means following the problem (Brewer, 1999), which in turn forces the scale to be framed in ways central to the problem. However, in practice, it is not clear that our disciplines prepare us well for allowing our scales to be framed in ways other than those disciplines.

An important but less recognised aspect of interdisciplinary work stems from the attitudes and feelings of the co-researchers. There are two aspects to this issue which may be related, but each is important in their own right. Firstly, lack of respect between physical and social scientists is frequently mentioned (e.g. Bruce *et al.*, 2004; Evans and Marvin, 2004; Lowe and Phillipson, 2009). This is not just about the worth of interdisciplinary research but can run deeper and can be founded in a lack of respect for each discipline, particularly the approaches and methods for undertaking research. Secondly, when disciplinary divides are transcended the novelty and quality of the research produced is often questioned. This in turn leads to interdisciplinary research being regarded as of lower status; and as a consequence professional status and promotion may be adversely affected (Brewer, 1999). Complementarity and cooperation, rather than competitiveness, may help in overcoming the negative effects of interdisciplinarity, and shared language has an important role to play here, both in theory and in practice (Bracken and Oughton, 2006). Once different types of knowledge are seen as embedded in different cultural contexts and there is mutual respect between specialisms, important lessons can be learnt and a much more fruitful collaboration instigated. More than this, the success of interdisciplinary research depends on nourishment drawn from shared disciplinary competence (Hansson, 1999), which in theory should remove any hierarchical value between different subjects. However in practice this may be difficult to attain and is strongly determined by the personal characteristics and interrelationships of those involved in undertaking the research.

The role of language in conducting interdisciplinary research

To try to improve practice in interdisciplinary research, previous work has tried to understand the ways in which we use language and how it may help or hinder our understanding of each other and our respective disciplines. This work has been concerned with the contemporary, multiple meanings and uses of words in practice. Language may determine the positionality of the researcher, the way in which the research question is framed, the translation of the 'field' to the academy, and the development of the theoretical context (see for example, Quinn and Holland, 1987; Mirowski, 1994; Pryke *et al.*, 2003). In the course of my work there have been occasional but important instances where the meaning or intent of the speaker has not been understood by the listener. This led Bracken and Oughton (2006) to identify three distinct issues: registers of speech termed *dialects* and *metaphor* and the process of *articulation*.

Dialects

Dialects represent the difference between everyday use of a word and specialist, disciplinary use of a word (Wear, 1999; Bracken and Oughton, 2006). Dialects are also produced by the same word having slightly different meanings within different disciplines (Bruce *et al.*, 2004), which may also be different from their everyday meaning. Words which are in everyday use tend to be those that cause the most difficulty for the unwary interdisciplinary scientist. These misunderstandings may be exacerbated by the fact that academics, often trained to be articulate, are unlikely to question the meaning of a word with which they are already familiar. The conversation may be well developed before it becomes apparent that a particular word has a specific disciplinary interpretation as well as its everyday use. This situation is bound to lead to frustration for those coming together from different academic specialism's to undertake interdisciplinary research.

The word Bracken and Oughton (2009) used as an example is *dynamic*. Our analysis was of a discussion about processes and methodology that took place in the field between researchers drawn from both physical and social science. In a discussion of farmers' understanding of the landscape the physical scientist described the catchment response to rainfall as being a *dynamic* process. This scientist argued that farmers were likely to understand how processes varied between storms and at the annual timescales, but their understanding of detailed hydrological processes operating within a storm may be less good. The social scientists took for granted that farmers would understand hydrological processes through their experience of living on and working the land, but did not pick up the different time period implicitly implied by the physical scientists when they used the word dynamic; that is that hydrological processes may vary over minutes as well as longer periods. The physical scientists did also not think to highlight the time period about which they were thinking. This different emphasis of the time period associated with the word *dynamic* led to a series of heated debates about different stakeholders' knowledge of the landscape (Bracken and Oughton, 2006).

In this example *dynamic* has both everyday meanings and discipline specific meanings. The problem lay in the differences in the perceived time and spatial scales to which *dynamic* referred between disciplinary and normal use. To the physical geographer *dynamic* meant that stream discharge would be variable depending on the antecedent moisture conditions of the catchment over very short timescales of a few hours to a few days. The social scientists understood *dynamic* to mean relatively rapid changes but did not associate changes with a specific time period because they did not have the same disciplinary training on which to draw. This confusion could easily have been clarified on the spot had we recognised this as a *dialect* word. This example shows how we got to very different endpoints from a poor matching of understanding of one word. In the company of experts of the same discipline this misunderstanding would (probably) not have happened. In this instance the difference in language was a constraint because it led to the physical scientist feeling misunderstood. However, at other times differences in language can be an opportunity and enable multiple interpretations which offer the potential to enrich other's framing of an idea or interdisciplinary research.

Metaphor

The second aspect of language is *metaphor*. The use of metaphors is common in our everyday discourse and is embedded in our language; we rarely think about them or are aware that we use them. When working within disciplinary teams we tend to have common metaphors on which we draw to express and explain ideas. For good interdisciplinary practice we need to be aware of the times at which we move into separated speech communities and when the form of metaphor being used may be misinterpreted

(Bracken and Oughton, 2006). The metaphor that I have previously explored is *mapping*. The context of the discussion was the development of the analytical framework to underpin a research study. Understanding and use of metaphor did not bring about the type of discipline related disagreement that *dialects* had generated. In this case, the project team were conscious of the role that *mapping* played as a metaphor. We did not intend to use the term in the sense of either 'relational hieroglyphics to represent the landscape' or as a diagrammatic systems framework as used by both physical and human geographers. We were seeking a framework that would allow us to relate differently conceived social relations and embed them within the physical landscape, to explore the complex interrelationships ongoing within rural areas between different groupings of humans, but also between humans and the landscape. We worked on this idea together and the metaphor *mapping* provided us with a name for this activity. In this way we could achieve agreement because we were using a relatively empty metaphor. This conjures up an image of a multilayered, complex web which is both affected by and affects everyday life. *Mapping* is the systematic description of the processes involved. This is also a very different interpretation of *mapping* from that used in everyday language, and is therefore a dialect word as well. Its use was clear to us because we developed it explicitly to meet our needs in this research project. However, were we to simply refer to our work in discussion with others as 'mapping catchment interactions' a completely different product could be imagined.

Articulation

This aspect of language differs from the first two in that it is a process rather than a register of speech. *Articulation* as described by Ramadier (2004) involves deconstructing one's own disciplinary knowledge in conjunction with those of other disciplines in order to understand the building blocks and thereby reconstruct a common understanding. We found the idea of *articulation* particularly stimulating and an accurate description of the very active discussions involved in framing and undertaking interdisciplinary research.

The context for our discussion was the first meeting to discuss the development of a research proposal. The physical scientist started to talk about her work in a river catchment. One of the social scientists asked what *catchment* meant. The physical scientist described a *catchment* as the area of land defined by the watershed (drainage boundaries) of a particular river. The social scientists were concerned that it had little meaning as a boundary for social and economic processes related to the physical landscape. Furthermore the economist and human geographer had different conceptions of what these might be. These slightly different definitions of *catchment* serve to highlight two alternative starting points in terms of thinking about the landscape, related to disciplinary backgrounds. Breaking these down and developing a shared, common understanding was a more powerful platform from which to develop a project.

Through the process of articulation we moved much closer to the crux of the problem that we wished to explore but, as a by-product, gained a deeper insight into what interdisciplinarity really meant. One aspect of this progress was to define a whole that was greater than the sum of its parts. A second aspect emerged through the process. Each member of the team was constantly tested in their assumptions and perceptions, and whilst the process was difficult and time consuming it was rewarding and resulted in a much stronger basis from which to develop research (Bracken and Oughton, 2006). It must also be noted that discussion, debate and disagreement can also produce new knowledge, rather than just simply translating knowledge onto a single, agreed and/or negotiated axis (Whatmore, 2009).

Reflecting on language and listening

Initially, explaining disciplinary knowledge to other academics may appear easy, whereas in reality communicating the formal knowledge is easy, but the more subtle intricacies and detailed theoretical context are much more difficult to communicate (Bracken and Oughton, 2006). A problem for the person speaking is that they need to be able to imagine their knowledge outside of their usual working context and practice in order to be able to communicate effectively. This is a point at which it is particularly important to recognise different cultural dialects and metaphors. But because of their different trainings they do not have the subtleties of discipline specific language and metaphor. Each contributor in the discussion may believe that they have knowledge of the others' understanding, although this will not necessarily be so. Only through a process of active and engaged articulation can these shortfalls come to light (Bracken and Oughton, 2006; Whatmore, 2009).

Academics bring with them a commitment to discovery and are active learners. A feature that is likely to arise when talking to other academics is a feeling of being challenged by experts in other fields. This can generate frustration, defensiveness, or even feelings of superiority, as it becomes difficult for the researcher to make themselves understood. Very quickly they may even feel that their work is being undervalued by others. On the other hand there is a danger of appearing cavalier about other peoples' knowledge. Emotions such as these can make the research team vulnerable to disciplinary competitiveness, which can immediately limit the effectiveness of the interdisciplinary research. Yet, the process of challenging others and of being challenged oneself is the essence of academic practice and often most progress is made when we discover that something we thought to be true is not. However, the arena in which the process of challenge takes place is important in either enabling new knowledge to be produced, or in closing down conversations. Within an arena where mutual respect exists between researchers mutual trust can be built which enables researchers to explore boundaries between different perspectives, specialisms, methods and approaches, which in turn is more likely to support the production of new knowledge.

The micro-politics within the group can also influence effective communication. Personal characteristics have a huge role to play in enabling successful interdisciplinary projects (Bruce *et al.*, 2004), but in my own interdisciplinarity research there have been a number of occasions when conversations between different disciplinary specialists could have been supported and developed in more depth when there were more than one representative of a discipline present to engage in the conversation. Two people, with a shared disciplinary background, can bring different perspectives, draw on alternative language and examples to assist interdisciplinary understanding and offer support. This may help to prevent any frustration and/or defensiveness developing from being misunderstood or feeling that you cannot communicate effectively. The cumulative effect of frustrated and ineffective communication and a feeling of disciplinary vulnerability may be disastrous. If the negative emotions become too great it may seem easier to just walk away, and engage in parallel rather than true interdisciplinarity research.

Framing

Framing encompasses the processes of identifying and bounding the area of research (Miller, 2000). The use and role of language can be important in developing the frame by making sure people from different disciplinary backgrounds understand each other, but by also enabling new knowledge to be produced from which to develop a frame. There are many definitions and descriptions of the process of framing, for example: 'frames function to organize experience and guide action' (Snow *et al.*, 1986:464); 'a sense making device, adding meaning to previously confusing or less meaningful situations or domain' (Weick, in Dewulf *et al.*, 2007) and 'the different ways in which actors make sense of specific issues by selecting the relevant aspects, connecting them into a sensible whole, and delineating its boundaries' (Dewulf *et al.*, 2004:178). Few studies have explored practices of framing, although there are some guidelines for good practice (see for example Dewulf *et al.*, 2004; Mostert and Raadgever, 2008). Distinctions have been made between framing and sense making. Fiss and Hirsch (2005) suggested that 'sense making stresses the internal, self-conscious process of developing a coherent account of what is going on, while framing emphasises the external, strategic process of creating specific meaning in line with political interest' (Fiss and Hirsch, 2005; 31). Oughton and Bracken (2009) do not distinguish between the two because they see both as being an integral part of the framing process.

There is a considerable amount to be learned about the negotiations behind successful framings and the factors that affect them that may be of use to the researcher embarking on interdisciplinary risk research. Within traditional disciplinary ways of working the framing of the research question often arises directly from previous research or is left as implicit. The process of framing the research is not explicitly defined because it is accepted, or habitual, within a school of thinking which shares an ontology and epistemology. Hence the research process tends to follow an established pattern and is concerned with limiting a distinct experiment or investigation rather than placing it in a context (Oughton and Bracken, 2009). Explicit negotiations associated with framing in this context need not be extensive and may be minimal. This is not to say that framing undermines the integrity of the research but it may be that it is just not necessary. In the case of an interdisciplinary project none of this process can be taken for granted. Extra negotiation and uncertainties surround working in a research project that takes place outside and across disciplinary boundaries. By exploring the activity of framing across a number of different projects it is possible to identify a range of features that support and shape the successful initiation and continuity of the research process (see Oughton and Bracken (2009) for more details).

Oughton and Bracken (2009) proposed two core elements to framing: identifying the research focus and determining the common objects with respect to scale, place and time; and identifying the research team. The team may exist before the idea or they may be brought together in line with the research question. The purist may argue that the people are not part of the framing, but because they are so intimately connected to the framing processes we have chosen to include this as a core activity. Hence, researchers affect the research question and the research question determines who is drawn in as a researcher. Many of the ways and processes of developing interdisciplinary research projects are no different from the initiation of ideas in disciplinary projects, but what is very different is the negotiation that is involved in developing the frame. This is a very practical process where the machineries of knowledge construction are really brought to the fore (Knorr-Cetina, 1999). Oughton and Bracken (2009) interviewed 12 people leading interdisciplinary research projects. One aspect of the process that three of the interviewees commented on explicitly was that the iterative and negotiated natures of developing the research project which may well mean withdrawing a particular contribution, take or person from the project should their input no longer seem relevant: in this case, the frame. The issue of writing yourself out of interdisciplinary research has also been raised by Lane *et al.*, (2006) as a necessary risk of pursuing interdisciplinary research.

There is also a third element to developing the frame of research, especially relevant to undertaking research around risk; the problem itself and letting the problem exist as a frame. This element of framing is therefore about letting those who live with risk challenge the ways that we frame risk, similarly to Lane (2008), where he allows fish to tell us what makes them hydrologically happy, rather than relying on what hydrologists think make fish hydrologically happy. Allowing those to speak back who matter might not simply involve 'normal conversations' such as those who live with flooding, but conversations with a much wider set of 'things', both animate and inanimate, that constitute and formulate the problem that is of interest.

The external research environment

There are three closely related aspects of the external research environment that influence framing: the political economy of research funding, the desire of researchers to feed into the policy process to initiate social change, and people experiencing the problem. As one interviewee argued in the Oughton and Bracken (2009) project, the research environment established by calls for context driven research means that the framing is established prior to the research call. One pragmatic response to funding calls is to take an idea off the shelf and add to it to meet the particulars of the research call. Although it was acknowledged by interviewees that this was not likely to be the most successful way of working it can be useful for developing joint studentships, evolving failed applications for research funding, and using scoping studies. Interrogating the funding environment is the first step for some. The need for policy relevance was also noted, not just as an important aspect of funding applications, but also as a significant driver to research from the perspective of researcher motivation – 'unless work is policy relevant it won't be useful'.

For all the projects carried out by the respondents in work reported by Oughton and Bracken (2009) there was a strong practical/political element. The problem focus draws attention to a second important range of factors affecting framing, the personal motivations and experience of the researchers. The shifting institutional context has opened up opportunities for researchers whose personal motivations are focused on wanting to 'make a difference' or 'give something back'. So the 'research problem' becomes clearly placed in its social context and needs to be defined/framed in such a way that it can be solved. Although applied research is not always of the highest quality; it may not need to be, to be useful and 'make a difference'.

In the current funding crisis surrounding higher education, universities have also been trying to influence the type of research undertaken. The range and number of university institutes is on the rise, with research proposals being encouraged to map on to these institutes, often with an element of interdisciplinary research. This type of environment moves the focus from the government steering mechanisms for research through research programmes such as those offered by RCUK (Lowe and Phillipson, 2006), to more everyday areas of engagement within researchers' own institutions.

Reflexive processes and negotiations

Institutional context, personal values and expertise bring people to a meeting of ideas from which the iterative and dynamic process of outlining the research problem begins. This emphasises the key point about framing that it is an active and dynamic process, an interaction (Bouwen, 1998; Gray, 1989; Gray and Donnellon, 1990; Drake and Donohue, 1996; Oughton and Bracken, 2009). Actors bring to the discussion different perspectives related to their values, training, interests and political stance. The most obvious being their disciplinary background, which influences training and experience (Bouwen *et al.*, 1999; Dewulf *et al.*, 2004; Donaldson *et al.*, 2005). Differences in the cultures of natural science training, as opposed to social science training, were mentioned as significant in a number of ways in the interviews conducted by Oughton and Bracken (2009). Questions may be formulated quite differently within these different cultures. Wider personal interests play into this meeting of ideas: the different priorities held by investigators, the type of solutions that they are looking for, how their current research maps onto the subject of investigation, and the social networks to which they belong are all significant (Bouwen *et al.*, 1999; Dewulf *et al.*, 2004; Gray and Donnellon, 1990).

The emerging outcomes of the research may also feed back to reframe a project. For example Pahl-Wostl (2002) sees framing as providing important elements of social learning such as, constructing shared problem definition, building trust, critical self-reflection and recognising different perspectives – all may feed back to initiate change. New interpretations offered by one party can provoke reinterpretations of a situation by the others (Gray and Donnellon, 1990). Drake and Donohue (1996) take this further by suggesting that each move in the discussion frames the issue in a specific way and propose the frame as an 'interaction mode' which can be accepted or rejected by the other participants through respectively maintaining or altering the frame in response. Parties to the discussion may of course include stakeholders from beyond the academic community. This is vital third element of framing where the problem is allowed to do the framing itself, and finding mechanisms that do this.

Research funding, inter-disciplinary negotiation and location are all limited in how far they can go without finding methods that allow the problem 'to speak out'.

What came out strongly from those interviewed was not just the amount and intricacy of debate that was necessary at the outset of establishing an interdisciplinary research project, but the continued negotiation that was demanded throughout the lifetime of the research (Oughton and Bracken, 2009). This is also because the act of researching, or coming to know, a particular framing becomes, in itself, something that unsettles that framing in a similar way which can be termed 'knowledge shifts'; this draws parallels with Kuhn's paradigm shifts. It also emerged that researchers found it important to recognise that interdisciplinary projects are iterative because they are drawing upon a number of different disciplines and that the research may be particularly complex and liable to fail in some aspect. You therefore have to be ready to change to ensure an outcome, especially with the involvement of stakeholders who are often looking to the research to solve immediate problems. Although we talk about a starting point, the process continues as the research project develops. It doesn't end after research funding is secured or after the work starts, but is an active part of the research process (Oughton and Bracken, 2009). Researchers found it important to understand what each other were saying and meaning to be able to negotiate the developing frame; thus the explicit use of language was important.

Finding the common focus

Finding the common focus is a process of careful inclusion, or closure, which separates the research from the wider environment, i.e. what is to be investigated and what is not. An important aspect of framing at this point is how to determine the closely related question of the spatial, temporal and institutional limits to the research. Place may be a key concept in bringing these together, particularly the choice of case studies. The choice of case studies represents the empirical outcome of deciding the scope of the research and negotiating scale is a necessary part of this process. But, negotiating scale is also a part of choosing the analytical approach to be adopted in the work because scale may lead to certain methodologies being relevant or not. There is a strongly pragmatic element in the choice of scale. The research needs to be determined broadly enough to capture the richness and complexity of the problem, but at the same time to keep the research project manageable. The relationship between choices of temporal and spatial scale and approach can be central to developing the frame through the processes of articulation, negotiation and new knowledge production (Bracken and Oughton, 2006; Oughton and Bracken, 2009; Whatmore, 2009). Once again, the problem has an important role in constraining the scales that matter and hence the disciplines that are needed.

Case studies are very important for bringing together interdisciplinary practices and this space is often a geographical place where people already have built up a history of working. Places can also be 'truth spots' that challenge the way we have constructed the world (i.e. our framings) in particular disciplines and so very valuable in unsettling those framings and allowing new and more interdisciplinary framings to come about (Gieryn, 1999). Choosing case studies also offers the opportunity to visit new places and create new spaces for thinking about ideas. Therefore, in some frames, researchers suggest case studies and locations with which they are familiar. Negotiation is therefore immediately focused on certain scales and environmental processes and possibly short circuits some interesting discussions. In other instances, the process of framing develops new case studies which ensured decisions about scale and processes were made explicitly rather than implicitly in framing. It is important to be explicit in interdisciplinary research projects because there is a need to relate different epistemologies and methods, which is often not necessary in projects with a disciplinary focus (Oughton and Bracken, 2009).

Language and framing in relation to risk research

Risk is a focal topic of many disciplines, professional activities and practical actions around natural hazards, technological threats, health impacts, crime, terrorism and pollution. Most scientific disciplines provide risk analysts with specialised competence needed in the study of one or other type of risk. Different disciplines also tend to offer their own overarching approaches to understanding and exploring risk. Some of these approaches are interdisciplinary, or give rise to cooperations between disciplines. Many also draw heavily on stakeholders. One key feature of risk research that ensures interdisciplinarity is that risk is strongly connected with normative issues and investigations are often motivated and influenced by debates on how societies should deal with risks. So can experience of undertaking interdisciplinary research assist those working on interdisciplinary projects around risk?

One of the most relevant aspects of the discussions above to undertaking risk research is the role of the problem in developing the frame for the research. This is vital in developing interdisciplinary projects around risk that are founded in 'wanting to be useful' and assist is making a difference to people affected by or living with risk. This is highlighted by the wealth of risk research that has explored the notion that interventions change the spatial distribution of risk which, with time, is consequential for the process of change (Lane *et al.*, 2011). The nature of an intervention will be very different depending on the perspective from which the risk (or problem) is approached. Mustafa (2005) and Lane *et al.*, (2011) highlight this by considering the range of approaches taken to flood management, in particular science based management decisions, versus experiential approaches that are grounded in the understandings of those who live with risk. The interpretation of the risk, perceived cause and preferred solution will all vary depending on the personal experience of risk, or involvement in managing risk. This is important in developing the frame and should be enabled to inform the perspective of the project, rather than solely identifying the focus and the team.

Lane *et al.*, (2011) also demonstrate that a suite of scientific practices exists in terms of flood management that might support interventions for risk management, but that this suite has to be constrained by policy objectives: it is policy objectives that constrain scientific practice; rather than scientific knowledge that constrains policy. I have found parallels in my own interdisciplinary work on being involved in making decisions around catchment management to protect the endangered freshwater pearl mussel in the UK. Reflecting on being involved in a committee formed to try to protect and enhance pearl mussel populations underlines the selective use of expertise and evidence which may provide a number of ways of achieving a given objective. Furthermore, the presence of uncertainty, both with respect to scientific findings and the social behaviour of human beings means that management decisions should not be viewed as rigid or the only version of 'the truth'. In the planning cycle there are multiple points where equally compelling choices are available. That is, there may be more than one narrative describing the route to achieving a particular outcome. Different bundles of evidence and/or expertise could produce the same or different conclusions. It is important to recognise and reflect on the suite of practices of management interventions which exists to inform and shape the frame of a particular project. This is especially important in the continual reworking of the frame that is ongoing in interdisciplinary projects.

In letting the role of the problem take centre stage in developing the frame of research around risk, stakeholders come to the fore in the processes of articulation and negotiation. Key individuals can play multiple roles in driving and shaping a project. Current risk research often incorporates models that identify vulnerable conditions that affect people, but also considers societal resistance and resilience (Cutter *et al.*, 2003; Wisner *et al.*, 2004). 'Non certified' (Collins and Evans, 2002) experts have a vital role in providing experiential knowledge and perspectives on this research but also in developing new ways of thinking and engaging with risk. Certified experts and professionals working around managing risk can also be key intermediaries in developing and undertaking research.

In my own work around managing risk to biodiversity I have come across a key professional who was pivotal in creating a strong institutional context for action. This professional was both a facilitator and broker and in the commissioning and use of evidence to support management changes he was both translator and facilitator. There is a significant difference between each of these functions, which may be supportive of each other, but are quite distinct and require different knowledge and expertise. The role of translator is particularly significant for undertaking interdisciplinary research around risk; the communication between expertises grounded in different disciplines is vital and sensitive and it would be unwise to make an unquestioned assumption that what one person says is understood to mean the same thing by another (Bracken and Oughton, 2006). By working on practical problems risk researchers have been good at reducing tensions between theory and practice (Cutter, 1993; Tobin and Montz, 1997; Montz and Tobin, 2011). Professionals and decision makers are thus ideally placed to inform the frame. However, they are not the only individuals who need to do this. Lane (this volume) outlines the ethical case for a more participatory approach to the framing process.

The role of place may also be more relevant for developing the frame in risk research than other interdisciplinary endeavours. This is implicit in 'letting the problem determine the frame' but should also be explored explicitly. Parallels can be drawn with the role of case studies highlighted by Oughton and Bracken (2009). Case studies tend to be chosen due to long term allegiances with an area and the increased depth of understanding that can be developed by repeated and related investigations. This may also be true for the role of place in developing the frame around risk research. However, place has a stronger role in framing risk research since it also determines the risk being explored, the key attributes that make the risk come to the fore as an avenue for research and most importantly the interdisciplinary nature of this risk which will be dependent on the risk but also social characteristics of the place. In this way particularly vulnerable regions and/or nations can act as entry points for both understanding and addressing the processes that cause and exacerbate vulnerability (Moss *et al.*, 1999; Brooks and Adger, 2003; O'Brien *et al.*, 2004; Brooks *et al.*, 2005).

Conclusions

Micro-politics play a significant role in the ability to practice good interdisciplinary work and this is reflected in the ways in which certain kinds of language become used and in terms of how projects become framed.

The notional hierarchies of disciplines, the personal ambitions and competitiveness of colleagues, not to mention the implicit and longstanding issues of power surrounding gender relations, all play an important part in determining interactions within a research team. Understanding the role of language in interdisciplinarity is not going to solve these problems – it does however offer a route to making them visible. Sharing and exploring a speech community involves transparency. There is nothing to hide behind if you are sincerely translating your work for others and simultaneously are engaged in actively listening to their contributions. This process involves becoming vulnerable and thus open to the misuse of power. Good interdisciplinary work is not therefore possible without mutual trust and respect. Recognising the roles of dialects, metaphor and articulation is a move towards creating a new space for interdisciplinary intellectual engagement. Related to this crucial use of language is the way in which we frame interdisciplinary research.

Individuals tend to have a strong predisposition to work across disciplinary boundaries and openness to collaboration with others. Successful projects are able to identify and support the processes that allow the communication and negotiation that is necessary, not just for the initial framing of a research funding proposal but are able to maintain negotiation. Self awareness and continual reflexivity and a willingness to be questioned by others are essential to this process. It is vital to acknowledge the continued role of reframing. This is especially important for projects collecting data across the social and natural sciences because it is difficult to map different epistemologies and methods successfully. Sub-projects which may encounter problems when underway should therefore not be written off, but negotiated to reframe them to find a way through the difficulties. Although hard, this is rewarding and may lead to new and exciting knowledge. What is required for successful framing of a research project is the coordination of the different perspectives that makes a problem appear and thus become amenable to research.

Understanding the practices of working in interdisciplinary teams is vital for successful research around risk as this type of research has become more overtly interdisciplinary itself. As well as the issues raised around language and framing in undertaking risk research the role of the problem in developing the frame should be explicitly recognised. Indeed it should be encouraged as a route to enabling researchers with different disciplinary backgrounds to come together and through the processes of articulation and negotiation be able to develop mutual respect and develop new knowledge to produce successfully funded projects on one hand, but to also ensure enjoyable research experiences with quality outputs and effective and useful suggestions for managing real work problems.

References

Balsiger, P.W. (2004) Supradisciplinary research practices: history, objectives and rationale. *Futures*, **36**, 407–421.

Bolland, J.D., Bracken, L.J., Martin, R. and Lucas, M.C. (2010) A protocol for stocking hatchery reared juvenile endangered freshwater pearl mussel Margaritifera margaritifera. *Aquatic Conservation: Marine and Freshwater Ecosystems*, **20**, 695–704.

Bouwen, R. (1998) Relational construction of meaning in emerging organizational contexts. *European Journal of Work and Organizational Psychology*, **7**(3), 299–319.

Bouwen, R., Craps, M. and Santos, E. (1999) Multi-party collaboration: Building generative knowledge and developing relationships among unequal partners in local community projects in Ecuador. *Concepts and Transformation*, **4**(2), 133–151.

Bracken, L.J., Cox, N.J. and Shannon, J. (2008) The relationship between rainfall inputs and flood generation in south-east Spain. *Hydrological Processes*, **22**(5) 683–696.

Bracken, L.J. and Kirkby, M.J. (2005) Differences in Hillslope runoff and sediment transport rates within two semi-arid catchments in south-east Spain. *Geomorphology*, **68** 183–200.

Bracken, L.J. and Oughton, E.A. (2006) 'What do you mean?' The importance of language in developing interdisciplinary research. *Trans Inst Br Geogr NS*, **31**, 371–382.

Bracken, L.J. and Oughton, E.A. (in press) Making sense of policy: The creative uses of evidence in managing freshwater environments. *Environmental Science and Policy*, Special Issue on Discourses and Expertises.

Brewer, G.D. (1999) The challenges of interdisciplinarity. *Policy Sciences*, **32**, 327–337.

Bruce, A., Lyall, C., Tait, J. and Williams, R. (2004) Interdisciplinary integration in Europe: the case of the Fifth Framework programme. *Futures*, **36**, 457–470.

Brooks, N. (2003) Vulnerability, risk and adaption: a conceptual framework. Working Paper 38, Tyndall Centre for Climate Change University of East Anglia: http://www.tyndall.ac.uk/

Brooks, N. and Adger, W.N. (2003) County level risk indicators from outcome data on climate related disasters: an exploration of the EM-DAT database *Ambio*.

Brooks, N., Adger, W.N. and Kelly, P.M. (2005) The determinants of vulnerability and adaptive capacity at the national level and the implications for adaption. *Global Environmental Change*, **15**, 151–163.

Buller, H. (2009) The lively process of interdisciplinarity. *Area* **41**(4), 395–403.

Carolan, M.S. (2008) The multidimensionality of environmental problems: The GMO controversy and the limits of scientific. *Environmental Values*, **17**(1), 67–82.

Collins, H.M. and Evans, R. (2002) The third wave of science studies. *Studies of Expertise and Experience: Social Studies of Science* **32**(2), 235–296.

Cutter, S.L. (1993) *Living with Risk: The geography of technological hazards*, Arnold, New York.

Dalgaard, T., Hutchings, N.J. and Porter, J.R. (2003) Agroecology, scaling and interdisciplinarity. *Agriculture, Ecosystems and Environment*, **100**, 39–51.

Dewulf, A., Craps, M. and Dercon, G. (2004) How issues get framed and reframed when different communities meet: A multi-level analysis of collaborative soil conservation initiative in the Ecuadorian Andes. *Journal of Community Appl Soc Psychol*, **14**, 177–192.

Dewulf, A., Francois, G., Pahl-Wostl, C. and Taillieu, T. (2007) A framing approach to cross-disciplinary research collaboration: experiments from a large-scale research project on adaptive water management. *Ecology and Society* **12**(2) 14 online URL: http://www.ecologyandscoiety.org/vol112/iss2/art/

Donaldson, A., Heathwaite, L., Lane, S., Ward, N. and Whatmore, S. (2005) Holistic approaches to understanding diffuse land management issues: A framework for interdisciplinary working. *Centre for Rural Economy Discussion Paper*, **2**, 1–14.

Donaldson, A., Ward, N. and Bradley, S. (2010) Mess among disciplines: interdisciplinarity in environmental research. *Environment and Planning A*, **42**(7), 1521–1536.

Downing, T.E., Butterfield, R., Cohen, S., Huq, S., Moss, R., Rahman, A., Sokona, Y. and Stephen, L. (2001) Vulnerability indices: Climate Change Impacts and Adaption UNEP. *Policy Series*, **3**, 1–91.

Drake, L.E. and Donohue, W.Q. (1996) Communicative framing theory in conflict resolution. *Communication Research*, **23**, 297–322.

Evans, R. and Marvin, S. (2004) Disciplining the Sustainable City: moving beyond science, technology or society, *The Resurgent City, Leverhulme International Symposium, LSE, http://www.lse.ac.uk/collections/resurgentCity/Papers/marvinevans.pdf*

Fiss, P.C. and Hirsch, P.M. (2005) The discourse of globalization: Framing and sense making of an emerging concept. *American Sociological Review*, **70**(1), 29–52.

Gieryn, T. (1999) *Cultural Boundaries of Science: Credibility on the Line*, University of Chicago Press, Chicago.

Gray, B. (1989) *Collaborating. Finding common ground for multiparty problems*, Jossey-Bass, San Francisco California USA.

Gray, B. and Donnellon, A. (1990) *An Interactive Theory of Reframing in Negotiation*, Unpublished manuscript, University College of Business Administration, Pennsylvania State.

Hansson, B. (1999) Interdisciplinarity: for what purpose?, *Policy Sciences*, **32**, 339–343.

Huby, M. and Adams, R. (2009) Interdisciplinarity and participatory approaches to environmental health. *Environment and Geochemical Health*, **31**, 219–226.

Karlqvist, A. (1999) Going beyond disciplines; the meaning of interdisciplinarity. *Policy Sciences*, **32**, 379–383.

Knorr Cetina, K.D. (1997) Culture in global knowledge societies: cultures and epistemic cultures. *Interdisciplinary Science Reviews*, **32**(4), 361–375.

Knorr Cetina, K.D. (1999) *Epistemic Cultures: How the Sciences Make Knowledge*, Harvard University Press, Cambridge.

Lane, S.N., Landstrom, C. and Whatmore, S. (2011) Imagining flood futures: risk assessment and management in practice. *Philosophical Transactions of the Royal Society A*, **369**, 1784–1806.

Lawrence, R.J. and Deprés, C. (1999) Futures of transdisciplinarity. *Futures*, **36**, 397–405.

Lowe, P. and Phillipson, J. (2009) Barriers to research collaboration across disciplines: scientific paradigms and institutional practices. *Environment and Planning A*, **41**, 1171–1184.

Medd, W. and Chappells, H. (2007) Drought, demand and the scale of resilience: challenges for interdisciplinarity in practice. *Interdisciplinary Science Reviews*, **32**, 233–248.

Miller, C.A. (2000) The dynamics of framing environmental values and policy: four models of societal processes. *Environmental Values*, **9**, 211–233.

Mirowski, P. (1994) *Natural Images in Economic Thought*, Cambridge University Press, New York.

Montz, B.E. and Tobin, G.A. (2011) Natural hazards: an evolving tradition in applied geography. *Applied Geography*, **31**, 1–4.

Moss, R.H., Brenkert, A. and Malone, E.L. (1999) Vulnerability to climate change: a quantitative approach, Pacific Northwest National Laboratory, Washington.

Mostert, E. and Raadgever, G.T. (2008) Seven rules for researchers to increase their impact on the policy process. *Hydrology and Earth System Sciences*, **12**, 1087–1096.

Mustafa, D. (2005) The production of an urban hazardscape in Pakistan: modernity, vulnerability and the range of choice. *Ann. Assoc. Am. Geogr.*, **95**, 566–586.

Nissani, M. (1997) Ten cheers for interdisciplinarity: The case for interdisciplinary knowledge and research. *Social Science Journal*, **32**, 1–14.

O'Brien, K.L., Leichenko, R., Kelkar, U., Venema, H. *et al.* (2004) Mapping vulnerability to multiple stressors: climate change and globalisation in India. *Global Environmental Change*, **14**, 303–313.

Oughton, E.A. and Bracken, L.J. (2009) Framing interdisciplinary research. *Area*, **41**(4), 385–394.

Pahl-Wostl, C. (2002) Towards sustainability in the water sector: the importance of human actors and processes of social learning. *Aquatic Sciences*, **64**, 394–411.

Pryke, M., Rose, G. and Whatmore, S. (2003) *Using Social Theory: Thinking Through Research*, Sage, London.

Quinn, N. and Holland, D. (1987) Culture and cognition, in *Cultural Models in Language and Thought* (eds D. Holland and N. Quinn), Cambridge University Press, New York, pp. 3–40.

Ramadier, T. (2004) Transdisciplinarity and its challenges: the case of urban studies. *Futures*, **36**, 423–439.

Smith, K. (1996) *Environmental Hazards*, Routledge, London.

Snow, D.A., Rochford, E.B., Worden, S.K. and Benford, R.D. (1986) Frame alignment processes, micro mobilization, and movement participation. *American Sociological Review*, **51**, 464–481.

Stenchion, P. (1997) Development and disaster management. *Australian Journal of Emergency Management*, **12**, 40–44.

Tobin, G.A. and Montz, B.E. (1997) *Natural Hazards: Integration and explanation*, Guilford Press, New York.

Wear, D.N. (1999) Challenges to interdisciplinary discourse. *Ecosystems*, **2**, 299–301.

Whatmore, S. (2009) Mapping knowledge controversies: science, democracy and the redistribution of expertise. *Progress in Human Geography*, 587–598.

CHAPTER 3
Religion and Disaster in Anthropological Research

Claudia Merli
Department of Anthropology, Durham University, Durham, UK

Abstract

This chapter summarises some of the major debates and contributions of anthropology to research on disasters and hazards and highlights recent trends in investigating the role of religion in the aftermath of catastrophic events. Anthropological studies of natural and man-made hazards have evidenced the importance of studying disasters as processes rather than events, as unfolding phenomena that need to be followed up in order to assess their real impact on local communities. An essential element of the recovery process from natural hazard is religion. Whether it is a visible component in the layout of religious-based NGOs in DRR or a rhetoric force present in local narratives, religion and cosmologies represent a driving aspect of communities' post-hazard adaptation. They cannot be reduced to appendixes to secular society because of their deeply political bearings. Whether mediating aid allocation, organising intervention, or commenting upon events individuals resorting to religious discourses take an active role in shaping the experience of people affected by hazard. Putting back religion into risk research enthuses cross-disciplinary reflections on the taken-for-granted allures of secular society, humanitarian intervention and scientific objectivity.

Introduction

Religious and cosmological ideas about environmental hazards and natural disasters that are often utilised by local cultures to explain these events, represent a challenge to contemporary risk research. Usually dismissed as mere narratives, at best relegated to the fields of theology and

Critical Risk Research: Practices, Politics and Ethics, First Edition.
Edited by Matthew Kearnes, Francisco Klauser and Stuart Lane.

anthropology, these discourses inform understandings of disaster and therefore also shape the ways in which cultures and societies adapt to these threats. It is for this reason that the ways in which disasters are understood in religious and cosmological terms should be the focus of interdisciplinary consideration in contemporary risk research. In order to provide an empirical starting point for this approach, in this chapter I review anthropological literature on disasters with specific reference to current research on the interplay between religious discourse and natural hazards. Rather than aiming to provide an exhaustive review of anthropological research on natural hazards, in this chapter I explore the dominant loci of academic debate in anthropological studies of disaster and natural hazards.[1]

The relative lack of empirical research on disasters in anthropology has been attributed to an epistemological preference in the discipline for investigations that favour forces of social change over and above those of sudden natural hazards (Torry, 1979b). However, disasters have gradually come to be regarded as totalising, multidimensional processes that are produced through a range of social, technological and natural factors and that also lead to long-term social change and adaptation (Oliver-Smith, 1996, 2002). It is in this sense therefore, that natural disasters, and the ways that societies adapt to them, constitute an ideal field of investigation for anthropologists, particularly from a comparative perspective. David Turton pointed out that the main contribution of anthropology to disaster studies is its focus on 'the cultural definition of, and response to, events in "nature"' (Turton, 1979: 533), making it clear that the difference between 'natural' and 'anthropogenic' disasters is itself hard to draw (Turton, 1979: 532). Disasters radicalise, and at the same time render problematic, the opposition between nature and culture, making of the emerging contradiction one of the signposts of catastrophes (Hoffman, 2002; Oliver-Smith, 2002; Signorelli, 1992). An archetypical expression of this radicalisation is to call into question the cosmological order, which is both challenged by natural disasters and utilised in articulating notions of responsible agency over these events. In the context of natural disasters, discourses that enable recourse to a divine (or supernatural) realm provide a source of explanation for these events.

Apart from its theoretical contribution, when it comes to research on disasters and hazards, anthropology also comes equipped with a unique methodology: '[d]own to earth, down to culture and society, ethnographic fieldwork greatly, indeed singularly, elucidates how disasters are constructed, what recovery entails and what the often enshrouded elements

---

[1] Excellent general reviews of the anthropological literature are Torry (1979b), Oliver-Smith (1996), Hoffman and Oliver-Smith (2002).

are that lead to a people's vulnerability in the first place' (Hoffman, 2010: 4). The groundedness of anthropological research is particularly evident in its focus on the 'community' as unit of analysis (Torry, 1979a), and returns in recent interdisciplinary collaborative research (Rigg *et al.*, 2008). Direct engagement and applied research on disaster are considered a privileged venue to produce theoretical change in the anthropological discipline. In the following sections of this chapter I focus on a series of specific temporal convergences of interest that help to explain the recent focus on religion and disaster in anthropological studies. In the mid-1950s pioneering works in the field were published as a series of special issues of *Human Organization*, the leading publication of the Society for Applied Anthropology, and the *Journal of Social Issues*. Between the late 1970s and early 2000s several monographs and edited volumes were also published. These works systematised the theoretical approach to disaster studies in anthropology, and had the effect of forming a sub-discipline of 'disaster anthropology'. In 2009–2010 recent special issues of anthropological journals and newsletters emphasise anthropological research on local post-hazard discourses and practices, in particular the place of religion in post-disaster social and political processes. This chronological unfolding of disaster as field of anthropological investigation goes hand in hand with the increasing relevance of critical and political economy approaches.

Applied Anthropology: disasters as social processes

In 1957 the journal *Human Organization*, one of the two journals related to the Society for Applied Anthropology (SfAA), published a special issue entitled 'Human adaptation to disaster'. The publication of this issue represented a culmination of a set of anthropological approaches to natural hazards, a field of studies which had catalysed academic attention only after the Second World War. The collection is a selection of papers presented at two meetings of the SfAA: the first was organised in Chicago in 1953 as a symposium on disaster (it is considered one of the earliest meetings entirely devoted to the topic), and the second was held in 1955 in Bloomington. Between these two meetings another special issue on disaster research was published in the *Journal of Social Issues* (see Chapman, 1954). Disaster research offered a unique venue of reflection in applied anthropology, a discipline intrinsically 'in action' that combines theoretical investigation and direct intervention in an unremitting dialogue, within and beyond anthropology. The editors of the *Human Organization* issue outlined the ambitions of this work in the following terms: 'Thus, in a very real sense, disaster studies should catalyze theoretical advances in other fields' (Demerath and Wallace, 1957: 1). One of the main points evident in the contributions to this volume is an integrated consideration of natural and man-made disasters. This can be considered the hallmark of the anthropological perspective on hazards and disasters, elaborated through the following decades in major anthropological works on the topic (Hoffman, 2002; Hoffman and Oliver-Smith, 1999; Oliver-Smith, 1977, 1996, 2002; Oliver-Smith and Hoffman, 2002).

The editors of the 1957 *Human Organization* special issue, Nicholas Demerath and Anthony Wallace, label the first period of research on disaster '"raw" empiricism.' What emerges in their publication is the centrality of the concept of 'system' in disaster research. For example they suggest that:

> Organization or system—we would use these terms interchangeably—changes are in response to forces originating either outside or inside "the system"... The *forces* that concern us here are, primarily, disasters, real or simulated, "natural" or man-made, comparatively sudden and surprising. The force tends to produce *stress*, defined as a condition of strain, contradiction or discrepancy between any of the parts or elements of a given system, or between that system and its field of environment. Stress, in turn, generates pressure for *change*; either change in internal *adjustment* processes or in the processes of external *adaptation* and defense. (Demerath and Wallace, 1957: 1)

The authors of the issue analyse three main systems of human behaviour: socio-cultural, perceptual and communicational, which are considered as in flux and mutable rather than stable (Demerath and Wallace, 1957).

The comment by the guest editors that '[f]ortunate and noteworthy, we think, is the fact that the papers do not point to a "disasterology"' (Demerath and Wallace, 1957: 1), is echoed nearly 50 years later in philosopher Adi Ophir's claim that there are no experts on disasters (Ophir, 2005, 2010). The field presents therefore a unique opportunity for collaborative and interdisciplinary research. In the closing comment of their introduction Demerath and Wallace (1957) also warn of the possible shortcomings related to approaching disaster research from a too mechanistic perspective:

> Also, with a more sophisticated approach to socio-cultural systems, it will be possible for us to become more exact in our statements of what a disaster is a disaster *to*. There is a tendency to speak of "a disaster" with only the impact agent and a statistic for casualties and physical destruction in mind, and to be very vague about the identity of the whole target. (p. 2)

Demerath and Wallace suggest that it is paramount to focus on the 'targets' (or victims) of hazards and their responses to the sudden collapse in social and cultural systems. In this early work, the term used to describe the conditions of shock, apathy or 'temporary paralysis' that follow a disaster is a 'disaster syndrome' (Wallace, 1957: 23). This terminology has some parallels with the more contemporary research on Post Traumatic Stress Disorder, a term that entered the psychiatric nosological system in 1980. However, this early socio-cultural research focused on concepts of 'adaptation', 'response', and 'adjustment' to change with the specific aim of providing a clear terminology for people working on the ground in disaster relief (Demerath, 1957). In contrast, in contemporary research on disaster we witness a shift of focus that privileges the psychologising vocabulary associated with the reconstitution of (or 'bouncing back to') an original state preceding devastation, that is 'resilience' (for an excellent critique of the concept see Manyena, 2006).[2]

The relevance of the applied perspective is reiterated in major works of later decades (Oliver-Smith and Hoffman, 2002), and addresses major issues of anthropological research, such as the connection between theory and practice.

Recovering from hazard: disaster anthropology

In the 1970s a more systematic approach to the anthropological study of disaster developed, along with a theoretical appraisal of the existing anthropological approaches (see Torry, 1978, 1979a, 1979b). The focus of this more recent work is twofold, focusing on the one hand on communities' capacity to maintain or re-establish stability (what Torry defines the 'homeostatic' approach, 1979b: 518) and on the other hand on the disaggregating forces that 'disrupt social stability' (defined by Torry as 'developmental approach', 1979b: 519). The formation of a 'disaster identity' – a concept used by Anthony Oliver-Smith to describe one of the processes taking place after the 1970 earthquake and landslide in Yungay (Peru) – is identified as 'a new factor in social differentiation' (Oliver-Smith, 1977: 6). The effects of individuals' categorization as either survivors (*sobrevivientes,* urban dwellers whose houses were smashed by the avalanche that buried their houses) or injured (*damnificados,* non-urban dwellers) were expressed in Yungay in terms of an entitlement to the distribution of aid and attached to notions of relative deservedness contra manipulation (Oliver-Smith, 1977). The task of delivering aid and support to the people affected by a disaster is accompanied by the phenomenon of 'convergence', in which

---

[2] The term is found in anthropological studies since 1970s, borrowed from ecology (Hollings, 1973 cit. in Torry, 1979b).

the local scene becomes congested by a plethora of outsider professionals often employed in reconstructive efforts and relief operations (Hoffman, 2010). In this context anthropologists noted that local knowledge and participation tended to be shattered to the advantage of decisions made by external actors (Barrios, 2010), and on the other hand it has since long become obvious that local skewed political powers manage to influence the allocation of international aid (for an excellent analysis of the aftermath of typhoon Pamela in Micronesia see Marshall, 1979). At times inequalities in distribution of aid are deeply etched along ethnic and religious lines (an example of discriminatory treatment of Muslim fisher folk in Sri Lanka following the 2004 tsunami is investigated by sociologist de Silva, 2009). In such situations anthropologists can take over the task of 'cultural translation' (Torry, 1978), mediating the interactions between local people and external individuals, who often resort to cultural and technical 'idioms' perceived as irreconcilable with those of the opposite party. In fact, a political economy approach to the study of bureaucratic organisation and disaster relief was introduced in the 1970s and evidenced the risk that disaster welfare could turn itself into a long-term risk (Torry, 1978, 1979b). The aim of this research approach is to unmask the 'culturally bound assumptions' leading disaster relief and aid allocation in extremely contextualised circumstances (Oliver-Smith and Hoffman, 2002).

Because of their processual nature, disasters are at best understood through long-term on-site ethnographic research and observation that go beyond the immediate effects of such events, and explore the range of local narratives, focusing on the variability of societal responses to disasters (Oliver-Smith and Hoffman, 2002). The process of environmental adaptation following a disaster is necessarily accompanied by an ideological readjustment, in the form of cultural adaptations, which include 'innovation and persistence in memory, cultural history, worldview, symbolism, social structural flexibility, religion, and the cautionary nature of folklore and fold tales' (Oliver-Smith and Hoffman, 2002: 9). Therefore, disasters unearth the relation between the ideological and the material (Oliver-Smith and Hoffman, 2002).

In sum I suggest that anthropological approaches to disasters function to deconstruct established dichotomies between: theory and practice; nature and culture; the ideological and the material. By corollary, and following Jürgen Habermas (2006), I also suggest that the division between the religious and the secular needs to be problematised particularly because of its blindness to the political role of religion in framing responses to disasters. For example, Habermas (2006) suggests that 'put differently, true belief is not only a doctrine, believed content, but a source of energy that the person who has a faith taps performatively and thus nurtures his or her entire life' (p. 8). In the kind of post-secular society that Habermas theorises, religion holds a central place in the public sphere. The research challenge is to investigate its relevance, particularly in times of crisis triggered by natural hazards.

Religion and disasters: an anthropological challenge

Anthropological research has focused on the way people re-think their place in the universe, the social system and ethical values that are articulated following a disaster, 'and include a deep delving into concepts of both social and cosmic justice, sin and retribution, causality, the relationship of the secular and the sacred, and the existence and nature of the divine' (Oliver-Smith, 1996: 308). The reoccurrence of disasters is also typically linked to cyclical punishments 'because of moral malfeasance' or corresponding to 'due' catastrophic events (Hoffman, 2002: 132). In these cyclical concepts of disasters, nature implies a notion of mastery, usually invested in God or other deities, which gives disasters a purpose, which transforms their destructive impact into a form of regeneration (Hoffman, 2002). Or, in other words, 'cosmologies and symbols readjust to clarify existence' (Hoffman, 2010: 4). These interpretations of catastrophes are not merely descriptions. Rather, they open up to the possibility of reformation (Journet, 2010) – they offer themselves to local contestations and also represent a field for the exercise of power (Oliver-Smith, 1996). The ways in which religious and cosmological concepts are tapped to render disasters intelligible should therefore be carefully considered to assess their impact on local capacities for adaptation and cooperation in post-disaster contexts.[3] For example, Dominic Johnson (2005) conducted a comparative statistical analysis of the data available from 186 communities (forming the Standard Cross-Cultural Sample [SCCS] selected by George Murdock and Douglas White in 1969) to test the hypothesis of a positive correlation between the belief in supernatural punishment (at the extreme variable meted out by 'high gods,' present and active in human affairs, and supportive of human morality) and the level of cooperation towards public goods in communities, in an evolutionary perspective. By examining a vast

---

[3] An article by Paton *et al.* (2008) is an example of a study which missed the point; researchers conducted a statistical survey to test levels of Collective Efficacy (CE) after the 2004 tsunami in Thailand in different communities. The team initially excluded ethnicity and religion because statistically 'insignificant.' In the discussion of the results they propose that religion may affect resilience but were unable to explain the mechanism through which this would be possible (Paton *et al.*, 2008: 117). The team was interdisciplinary and included psychologists and geologists/geophysicists.

corpus of literature Johnson shows that previous research that claimed to show that dispositions to cooperate were independent of religious beliefs are erroneous (p. 412). Rather he suggests that the relevance of belief holds both in relation to individual transgressions and following disastrous natural events; people think that the latter 'happened *for a reason*' (p. 415, original emphasis). He suggests:

> From there, it is a small step to assign the cause to some supernatural agency, given that such events apparently lie outside any human's ability to instigate them. (p. 415)

In the past two years special issues of academic journals and newsletters have been devoted to the study of disaster from an anthropological perspective. In several articles the relevance of religion in post-hazard social processes is clearly outlined; for example, in a special issue of *Religion* that was published in 2010, edited by French geographers Jean-Christophe Gaillard and Pauline Texier, on the theme of 'Religions, Natural Hazards, and Disasters.' In the same year, the journal *Terrain* devoted an issue edited by ethnologist Nicolas Journet to catastrophes. The two issues focused on several cases that demonstrate that the division between secular and religious worlds on the one hand, and natural and political worlds on the other, are neither assumed nor established definitively. Rather these divisions reflect a specific 'scientific' interpretation of disasters that does not correspond to the meanings attached to catastrophic events by local populations. The effect of religious interpretations in informing and framing local actions and responses to disasters therefore represents an important area for contemporary risk research.

The contributors to the special issue of *Religion* included anthropologists, theologians, and geographers. This was the first issue of a journal entirely devoted to the topic of religion and disasters and built on field-based evidence attesting to the resourcefulness of religion in the aftermath of natural hazard (Gaillard and Texier, 2010). The opening contribution by David Chester and Angus Duncan analysed the use of theodicy in the Christian tradition, highlighting the emergence of a discourse on the 'structural sinfulness' of unequal access to resources and aid (Chester and Duncan, 2010). It also addresses the issue that despite persistence of religious explanations in local populations, these discourses are seldom recorded, partly because of 'a reluctance – even an embarrassment – on the part of many victims to admit that they attempted to explain losses in religious terms when answering questionnaire surveys carried out by university trained scientists and social scientists' (Chester and Duncan, 2010: 87), and partly because academic and government reports also tend to expunge these discourses. The closing contribution to the issue penned by Ben Wisner

(Aon-Benfield Hazard Research Centre UCL) analyses the possibilities offered by religions as agents of Disaster Risk Reduction (DRR) but he downplays the relevance of understanding the religious discourses to promote DRR and mobilize faith communities (Wisner, 2010). The four articles contributed by anthropologists, all working in Southeast Asia, focus instead on the importance of grasping local discourses in order to highlight political and social processes of disasters (Dove, 2010; Lindberg-Falk, 2010; Merli, 2010; Schlehe, 2010). The special issue of *Terrain* (published by 'Maison des sciences de l'homme') was entitled 'Catastrophes', with contributions by ethnologists, anthropologists and historians. The issue does not focus explicitly on religion but the topic is central in four articles. It includes studies on disasters triggered by natural hazards (Brac de la Perrière, 2010; Journet, 2010; Quenet, 2010; Revet, 2010), as well as studies on 1997 avian flu pandemic and SARS in 2003 analysed in terms of biosecurity (Keck, 2010), the food crisis and new 'eating' programmes introduced following Katrina in New Orleans (Larchet 2010), on photography and catastrophes (Solomon-Godeau, 2010), and on the Shoah of gypsies (Stewart, 2010). I will not refer in detail to these interesting four latter works. The articles I look at closer span the continents, addressing disasters in Latin America (Venezuela), Europe (France), and Asia (Myanmar). The issue opens with an article by Nicolas Journet, 'Catastrophes et ordre du monde' (literally 'Catastrophes and world order').

The two special issues offer a range of complimentary themes. Three of the contributions in the *Religion* issue specifically address the interplay of religion and politics in the construction of local cosmologies (Dove, 2010; Merli, 2010; Schlehe, 2010) and one focuses on the bereavement and funeral rites following the 2004 tsunami in Thailand (Lindberg-Falk, 2010). Michael Dove portrays a link between macrocosm and microcosm in Java's people consideration of Mount Merapi (site of the spirit world) as mirroring Javanese society through the Yogyakarta *kraton* ('residence of the king'). The activity of the volcano is monitored by people who consider it a reflection of society's activity and perturbations. Dove applies the Foucauldian concept of a panopticist self-surveillance to explain this relation (Dove, 2010), showing 'that religious and cosmological beliefs have an under-appreciated relevance to the current interest in state surveillance' (Dove, 2010: 125). Political uses of earthquakes are also recorded in European history, for example in France where Louis XIII made use of the quake of 3 July 1618 to discredit the Protestants (Quenet, 2010). We can infer that a parallelism between administrative control of society and administrative control of nature developed in the West at the same time. The concept of risk, with its aura of control, management and probabilistic calculus (and anticipation) is not equivalent to catastrophe, and in this sense expertise does not overlap (Quenet, 2010: 12). For example, Quenet

(2010) opens his article by addressing the terminological confusion and interchangeability of the concepts 'risk', 'threat' and 'danger' (*risque, menace, danger*) often affecting the reception of historical works on the topic (2010). He shows that attitudes toward the earthquakes changed in France during the eighteenth century, to include administrative measures that accelerated 'the secularization of the view on nature' (p.19).[4] After the Lisbon earthquake of 1755 the secularization of natural catastrophes is enhanced by transforming 'the drama into a contingent event set in a specific time and space' (p. 21).

Different hazards are marked by alternative temporalities between, for example, the recurring hazards of landslides, fires and inundations and the more exceptional cases of earthquakes. The development of a 'local seismic culture' (*culture sismique locale*) and a local memory of catastrophes is often a feature of the ways in which these more exceptional events are understood and narrated in cultures (Quenet, 2010). Despite the secularisation of catastrophes, we witness a return 'toward symbolical, mythological, religious or even "archaic" explanations' (Revet, 2010: 44) of these exceptional events. Resort to theodicy to explain disasters surfaces in contemporary societies, and across different faiths, without undermining scientific explanations. In fact these alternative explanations are often engaged at the same time (Chester and Duncan, 2010; Merli, 2010; Schlehe, 2010). For example, in Southern Thailand Muslim and Buddhist responses to and explanations of the 2004 Indian Ocean Tsunami functioned to re-establish the boundaries between ethno-religious communities with a long history of sharing the same territory and history (Merli, 2010). The religious elites commenting upon respective religious ideas did this well aware of the common themes and differences. The definition of boundaries can be considered a political act, especially when theodicies were employed to identify communities that sinned versus pious communities (Merli, 2010). By setting the devastated villages as examples, moral discourses direct people's behaviours.

Catastrophes also bring the possibility of renewal and are associated with notions of new beginnings and transformation which often have moral overtones, connected to the reformation of conscience (Revet, 2010; Schlehe, 2010). At times this reformation is intended as a political one. For example, Schlehe (2010) analyses how the 2006 earthquake in Indonesia was considered to be the result of collective wrongdoings that elicited social conflicts and the way different local groups' political claims and dynamics were expressed. In an attempt to restore harmony between nature and society, people claimed that the Sultan was to blame for the

[4] Translation by the author.

earthquake because he had given up his traditional (cosmological) role of syncretistic Islam in favour of a more literalist and modernist version (Schlehe, 2010). Two other ethnographic examples demonstrate how the cosmological and the political levels can be interwoven. Revet's (2010) research on the massive mudslides that devastated Vargas (Venezuela) in 1999 demonstrates that as these events coincided with a national vote for the new Constitution the concurrence influenced the way this event was understood. This chronological convergence triggered a range of religious and political interpretations. For example, the Catholic Church, notably the archbishop of Caracas, accused the Chávez presidency of pride. Similarly, Protestant Evangelical Churches interpreted the events as a divinely authorised judgement against popular cults. Both of these religious and political interpretations were mirrored in local media reporting of the event. In both cases, nature is understood acting as an instrument of divine power or is replete with animistic agency and interiority and therefore takes revenge of humans. Analysing the connection between religious and political interpretations of disasters Revet (2010) suggests that:

> Prayer, respect for the environment or preventative measures have in a sense the same function: to restore in humans a sense of control on their destiny, a grasp on events, the capacity to act.[5] (p. 53)

In her analysis of the situation in Burma following the 2008 cyclone Nargis Brac de la Perrière (2010) makes a similar point. Like in the Venezuelan case the way that cyclone Nargis was interpreted was influenced by its coincidence with a referendum to ratify the new Constitution. The cyclone was interpreted as a punishment for the Junta that suppressed the protest of the monks in 2007. The interpretations of the cyclone in the weeks after the event pointed at signs that some monks and diviners had considered as announcing the catastrophe. The connection between this political and natural order is based on precedents in Burmese history, since 'extraordinary natural phenomena' have always announced the end of monarchic regimes. In this sense the catastrophe is understood as a 'karmic sanction, as it is reflected into the natural order, of mischief of the power in place' (p. 77).

A song composed after Nargis – containing the lyrics 'Pots de terre et vents de mer, secouez-vous!'[6] – expresses 'a cosmic unleashing mobilising people and the elements' (Brac de la Perrière, 2010: 79). The song

---

[5] 'La prière, le respect de l'environnement ou les mesures de prévention ont dans ce sens la même fonction : redonner aux hommes un sentiment de maîtrise de leur destin, une prise sur les événements, la capacité d'agir.'

[6] Pots of earth and sea winds, shake!

narrates the impact of Nargis by suggesting that the most afflicted parts of population on the delta were hit because of their exploitation of the marine resources. Similarly, a crisis in the Burmese fish market after the cyclone was accompanied by a story of a finger found inside a fish. Similar stories abounded after the 2004 tsunami in Thailand, when people stopped eating fish for fear of indirectly ingesting 'people' despite official contrary statements made by government ministries in Bangkok (Merli, 2005). Environmental encroachment – which is understood as a form of cosmological invasion –, overfishing, and stories hinting in a sort of both natural and cultural subversions at the spectre of anthropophagi in these kinds of narratives testify to the necessity to re-establish the respect of the boundaries between nature and culture.

At another level, one of the greatest challenges for communities experiencing an environmental disaster is probably the treatment of human corpses, including their formal identification and the performance of appropriate funerary rituals. For example, the treatment of the huge number of cadavers after cyclone Nargis involved the military destroying corpses using explosives. Collective rituals replaced individual funerals (Brac de la Perrière, 2010). In Thailand the impossibility of retrieving bodies led to funerals being conducted without bodies, a condition that many experienced as an 'ambiguous loss' (Lindberg-Falk, 2010).[7] In Burma and Thailand the presence of monks did not dispel the fear of the presence of malevolent ghosts. In Burma these spirits (*kyap*) could not reincarnate (Brac de la Perrière, 2010), while in Thailand survivors 'made merit' for the deceased by being ordered as monks (women as *mae chii*).[8] Monks played an essential role in Thailand to communicate between the two worlds and compared themselves to post offices to send off a person's *winjan* (spirit or soul) (Lindberg-Falk, 2010). In Burma instead apart from the presence of monks, to frighten the *kyap* taking human semblances people should shout the name of the popular organisation supporting the present Junta 'Djankaïnyé'. Westerners who brought aid were also identified as ghosts to be appeased (Brac de la Perrière, 2010).

---

[7] Lindberg-Falk calls this rite 'counterfeit funeral' ending in a 'counterfeit cremation.' I have some problem accepting both uses of the term counterfeit which usually denotes negative meanings; in fact, since there are both a funeral and a cremation they are neither 'simulated' nor 'falsified', they are just without a body or sometimes with a mock body in the coffin. The ritual explicates one of the aims it is supposed to absolve.

[8] In popular Buddhism the performance of good deeds, acts of devotion or renunciation (including temporary ordination of men as monks or of women as nuns) are not only considered important to accumulate karmic 'merit' for oneself (as in orthodox textual Buddhism) but also to transfer merits to the deceased (see Keyes, 1983; Obeyesekere, 1968).

Concluding remarks

The pivotal role played by religion after a disaster is increasingly becoming a concern in anthropological research. In a commentary following the 2010 earthquake in Haiti Hoffman points out that 'religions and ideologies change to incorporate events, invoke morality, deal with death and point fingers. Cosmologies and symbols readjust to clarify existence' (Hoffman, 2010: 4). She continues by highlighting the importance of anthropological research on disasters that incorporates a sensitivity to the role that religion plays in cultural interpretations and understandings of these events:

> Without question, anthropology as a social science is in a position to parse the totality of disasters in a way no other social science can. Only anthropology takes into its reckoning the three planes that interface in catastrophe: environmental, biological and sociocultural. The developmental and comparative perspectives of anthropological research also uniquely merge with disaster studies. (p. 4)

Addressing the religious dimensions of risk research therefore requires a thorough and cross-disciplinary re-theorisation of the often taken-for-granted distinctions between the religious and mythological aspects of disasters and the political and social factors that produce and intensify the effects of disasters. In particular this research suggests that the predominant secularism of contemporary risk research has the effect of downplaying the role of religion in framing cultural understandings of and responses to disasters. Religion also plays an important role in humanitarian intervention, especially since so many NGOs operating in DRR are faith-based. While this aspect can represent an asset to consolidate community resources and effort we should not dismiss the impact that these religious discourses have on the work of these agencies in local contexts (suffice to mention Haiti's 2010 earthquake and the plethora of religious NGOs and missions operating in the aftermath of the catastrophe). Religion in disasters and risk research also calls into question the assumed scientific objectivity of those assessing the effects of a disaster which have tended to neglect the forms of cultural adaptation to disasters limiting interventions to assessing physical, environmental and psychological damage. As several of the works briefly examined in this chapter suggest, religion is a primary aspect in post-disaster contexts, often in close relation to political change and claims. The holistic and multidimensional nature of disasters is probably one of the most relevant fields to implement a truly collaborative and interdisciplinary form of risk research, and one to which anthropology has much to contribute in terms of theory, methodology and engagement.

References

Barrios, R. (2010) Budgets, plans and politics: Questioning the role of expert knowledge in disaster reconstruction. *Anthropology News*, **51**(7), 7–8.

Brac de la Perrière, B. (2010) Le scrutin de Nargis: Le cyclone de 2008 en Birmanie [The Nargis election: The 2008 cyclone in Burma]. *Terrain*, **54**, 66–79.

Chapman, D.W. (ed.) (1954) Human behavior in disaster: A new field of social research. Special issue of *Journal of Social Issues*, **10**(3).

Chester, D.K., and Duncan, A.M. (2010) Responding to disasters within the Christian tradition, with reference to volcanic eruptions and earthquakes. *Religion*, **40**(2), 85–95.

de Silva, M.W.A. (2009) Ethnicity, politics and inequality: post-tsunami humanitarian aid delivery in Ampara District, Sri Lanka. *Disasters*, **33**(2), 253–273.

Demerath, N.J. (1957) Some general propositions: An interpretative summary. *Human Organization*, **16**(2), 28–29.

Demerath, N.J., and Wallace, A.F.C. (eds) (1957) Introduction to 'Human adaptation to disaster'. Special Issue of *Human Organization*, **16**(2), 1–2.

Gaillard, J.-C., and Texier, P. (2010) Religions, natural hazards, and disasters: An introduction. *Religion*, **40**(2), 81–84.

Habermas, J. (2006) Religion in the public sphere. *European Journal of Philosophy*, **14**(1), 1–25.

Hoffman, S.M. (2002) The monster and the mother: The symbolism of disaster, in *Catastrophe & Culture: The Anthropology of Disaster* (eds S.M. Hoffman and A. Oliver-Smith), School of American Research Press, Santa Fe and James Currey, Oxford, pp. 113–141.

Hoffman, S.M. (2010) Of increasing concern: Disaster and the field of anthropology *Anthropology News*, **51**(7), 3–4.

Hoffman, S.M., and Oliver-Smith, A. (1999). Anthropology and the angry earth: An overview, in *The Angry Earth: Disaster in Anthropological Perspective* (eds A. Oliver-Smith and S.M. Hoffman), Routledge, New York and London, pp. 1–16.

Hoffman, S.M., and Oliver-Smith, A. (eds) (2002) *Catastrophe & Culture: The Anthropology of Disaster*, School of American Research Press, Santa Fe and James Currey, Oxford.

Hollings, C.S. (1973) Resilience and stability of ecological systems. *Annual Review of Ecology and Systematics*, **4**, 1–23.

Johnson, D.D. P. (2005) God's punishment and public goods: A test of the supernatural punishment hypothesis in 186 world cultures. *Human Nature*, **16**(4), 410–446.

Journet, N. (2010) Catastrophes et ordre du monde. *Terrain*, **54**(1), 4–9.

Keck, F. (2010) Une sentinelle sanitaire aux frontières du vivant : Les experts de la grippe aviaire à Hong Kong [A health sentry at the frontiers of living: Bird flu experts in Hong Kong]. *Terrain*, **54**, 26–41.

Keyes, Ch.F. (1983) Merit-transference in the kammic theory of popular Theravada Buddhism, in *Karma: An Anthropological Inquiry* (eds Ch.F. Keyes and E.V. Daniel). Berkeley, CA and London, University of California Press, pp. 261–286.

Larchet, N. (2010) Réformer l'alimentation au lendemain de Katrina: La catastrophe comme prétexte à l'action [Diet reform following Katrina: Disaster as pretext for action]. *Terrain*, **54**, 80–99.

Lindberg-Falk, M. (2010) Recovery and Buddhist practices in the aftermath of the Tsunami in Southern Thailand. *Religion*, **40**(2), 96–103.

Manyena, S.B. (2006) The concept of resilience revisited. *Disasters*, **30**(4), 433–450.

Marshall, M. (1979) Natural and unnatural disaster in the Mortlock Islands of Micronesia. *Human Organization*, **38**(3), 265–272.

Merli, C. (2005) Religious interpretations of tsunami in Satun province, Southern Thailand: Reflections on ethnographic and visual materials. *Svensk Religionshistorisk Årsskrift*, **14**, 154–181.

Merli, C. (2010) Context-bound Islamic theodicies: The tsunami as supernatural retribution *versus* natural catastrophe in Southern Thailand. *Religion*, **40**(2), 104–111.

Obeyesekere, G. (1968) Theodicy, sin and salvation in a sociology of Buddhism, in *Dialectic in Practical Religion* (ed. E.R. Leach). Cambridge, Cambridge University Press, pp. 7–40.

Oliver-Smith, A. (1977) Disaster rehabilitation and social change in Yungai, Peru. *Human Organization*, **36**(1), 5–13.

Oliver-Smith, A. (1996) Anthropological Research on Hazards and Disasters. *Annual Review of Anthropology*, **25**, 303–328.

Oliver-Smith, A. (2002) Theorizing disasters: Nature, power, and culture, in *Catastrophe & Culture: The Anthropology of Disaster* (eds S.M. Hoffman and A. Oliver-Smith), School of American Research Press, Santa Fe and James Currey, Oxford, pp. 23–47.

Oliver-Smith, A. and Hoffman, S.M. (2002) Introduction: Why anthropologists should study disasters, in *Catastrophe & Culture: The Anthropology of Disaster* (eds S.M. Hoffman and A. Oliver-Smith), School of American Research Press, Santa Fe and James Currey, Oxford, pp. 3–22.

Ophir, A. (2005) *The Order of Evils. Toward an Ontology of Morals*, transl. by R. Mazali and H. Carel, Zone Books, New York.

Ophir, A. (2010) The politics of catastrophization: Emergency and exception, in *Contemporary States of Emergency: The Politics of Military and Humanitarian Interventions* (eds D. Fassin and M. Pandolfi), Zone Books, New York, pp. 59–88.

Paton, D., Gregg, C.E., Houghton, B.F., *et al.* (2008) The impact of the 2004 tsunami on coastal Thai communities: Assessing adaptive capacity. *Disasters*, **32**(1): 106–119.

Quenet, G. (2010) Fléaux de Dieu ou catastrophes naturelles? Les tremblements de terre en France à l'époque moderne [Plagues of God or natural catastrophes? Earthquakes in France in the modern period]. *Terrain*, **54**, 10–25.

Revet, S. (2010) Le sens du désastre : Les multiples interprétations d'une catastrophe ≪naturelle≫ au Venezuela [The meaning of disaster: The multiple interpretations of a "natural" catastrophe in Venezuela]. *Terrain*, **54**, 42–55.

Rigg, J., Grundy-Warr, C., Law, L. and Tan-Mullins, M. (2008) Grounding a natural disaster: Thailand and the 2004 tsunami. *Asia Pacific Viewpoint*, **49**(2), 137–154.

Schlehe, J. (2010) Anthropology of religion: Disasters and the representations of tradition and modernity. *Religion*, **40**(2), 112–120.

Signorelli, A. (1992) Catastrophes naturelles et réponses culturelles. *Terrain*, **19**, 147–158.

Solomon-Godeau, A. (2010) Photographier la catastrophe [Photograph the catastrophe]. *Terrain*, **54**, 56–65.

Stewart, M. (2010) Une catastrophe invisible: La Shoah des Tziganes [An invisible catastrophe : Shoah of the Gypsies]. *Terrain*, **54**, 100–121.

Torry, W.I. (1978) Bureaucracy, community, and natural disasters. *Human Organization*, **37**(3), 301–308.

Torry, W.I. (1979a) Anthropology and disaster research. *Disasters*, **3**, 43–52.

Torry, W.I. (1979b) Anthropological studies in hazardous environments: Past trends and new horizons [and Comments and Reply]. *Current Anthropology*, **20**(3), 517–540.

Turton, D. (1979) Comment to W. I. Torry's Anthropological studies in hazardous environments. *Current Anthropology*, **20**(3), 532–533.

Wallace, A.F.C. (1957) Mazeway disintegration: The individual's perception of socio-Cultural disorganization. *Human Organization*, **16**(2), 23–27.

Wisner, B. (2010) Untapped potential of the world's religious communities for disaster reduction in an age of accelerated climate change: An epilogue & prologue. *Religion*, **40**(2), 128–131.

CHAPTER 4
'Risk' in Field Research

Sarah R. Davies[1], *Brian R. Cook*[2] *& Katie J. Oven*[3]

[1]Center for Nanotechnology in Society, Arizona State University, Arizona, USA
[2]UNESCO Center for Water Law, Policy and Science, Dundee University, Dundee, UK
[3]Institute of Hazard, Risk and Resilience, Department of Geography, Durham University, Durham, UK

Introduction

We want to begin this chapter with an anecdote: one of the authors, as an eager graduate student, attends their first international academic conference, convened around European social theory and the politics of knowledge. Excited, nervous, and not quite sure what to expect, through the course of the three-day programme they find themselves struck, then surprised, and finally a little overwhelmed by what seem to be continual references to risk, even when the papers are on subjects as diverse as phone-based medical care and the Mexican psychological profession. In this context, at least – that of broad-based social research – risk appears ubiquitous.

It is certainly the case that risk runs throughout recent social theory: it is argued that we live in a risk society (Beck, 1992); are adrift in a fragmented and risky world in which we must continually reinvent ourselves (Giddens, 1991); have even become knowledgeable consumers of risks from extreme sports to genetically modified foods (Tulloch and Lupton, 2002). Risk has, since the 1960s, become an important analytical concept in both the social and natural sciences. While realist techno-scientific approaches remain dominant in many fields, socio-cultural conceptualisations – whether drawing on Douglas and Wildavsky's cultural theory, Beck's argument of transition from industrial modernity to a risk society, or Foucauldian-informed notions of governmentality – appear to be shaping thought across the social sciences (Lupton, 1999). Such ideas are, in turn, permeating risk research in general and hazard research in particular. There is a growing recognition of the need for socio-cultural approaches in understanding the risks associated with geophysical and

Critical Risk Research: Practices, Politics and Ethics, First Edition.
Edited by Matthew Kearnes, Francisco Klauser and Stuart Lane.

hydrometeorological phenomena (see, for example, Adams, 1995; Bankoff *et al.*, 2004; Wisner *et al.*, 2004).

Risk forms the basis of analyses of contemporary society, mapping new attitudes towards technoscience, new kinds of anxieties, and new ways of living. Current approaches to understanding risk highlight – correctly, we believe – the context-ladenness of risk-related research. Events such as the Chernobyl and Bhopal disasters have demonstrated the complexity and unpredictability of emergent risk within modern contexts. The notion of risk as a new global currency is clearly relevant, then, for empirical research. Equally clearly, as our somewhat naïve student's experience suggests, risk has been enthusiastically adopted by the research community. What is of interest to us, within this discussion, is the inter-relationship between risk as a conceptual and empirical device. How do we use 'risk' in our research? Is it a heuristic, a guide or something else? How does it structure our research processes, from design to fieldwork to the smoothly finished presentation of 'findings'? Or, in other words, how are theories of risk put to work, not just as 'background' for conference presentations, but in the messy business of conducting research and analysing data? These kinds of questions – which are not just about 'methodology', but rather the experience of embodied practices of research – tend to be treated with a degree of coyness in smoothed-over and tidied-up presentations and papers. The mess of field research and the (incomplete, creative, contingent) use of particular concepts are all too readily hidden by the tidy narratives of completed research (Law, 2004; Donaldson *et al.*, 2010).

As a way of engaging with the messiness of risk, then, we present a conversation – or at least the start of one – that involves discussion of three different ways of operationalising 'risk'. We explore how we put risk to work within empirical practices of research in contrasting geographical and disciplinary contexts. In doing so, we have been influenced by both the work of John Law (2004) on method and the 'mess' of inevitably incomplete research practices, and a number of scholars who have explored the practice of writing about academic research. We present our analysis, then, not in the traditionally sanitised and authoritative genre of the academy, but in the form of a conversation. We employ this approach as a way of engaging with our preconceptions and to see whether an alternate format might provoke innovative insights. We situate ourselves within the text, recounting our experiences and presenting a set of different and differing voices.

This format can be traced to a number of literatures, not least the brief flourishing – in the 1980s and 1990s – of 'new literary forms' within Science and Technology Studies (STS). Following their own analyses of the practices of science and the normative commitments its presentation entails, a number of scholars (including Michael Mulkay, Malcolm Ashmore and Greg Myers) suggested that 'new ways of organizing language' were required in order to represent the multiplicity of forms of social life studied. Mulkay (in an account of his own 'sociological pilgrimage') writes:

> In an attempt to address the self-referential nature of SSK's [Sociology of Scientific Knowledge] central claims and to display the ways in which analysts' claims are moulded by their use of specific textual forms, I began to employ multi-voice texts . . . Texts of this kind made it possible, I found, to replace the unitary, anonymous, socially removed authorial voice of conventional sociology with an interpretative interplay within the text (Mulkay, 1991: xvii).

Form, in other words, will affect content. Not surprisingly, the STS scholars who had demonstrated the value of reflexivity through analyses of the natural sciences (see, for instance, Latour and Woolgar, 1986) reapplied this analytical approach to themselves, arguing that the 'unitary, anonymous, socially removed authorial voice' which remains dominant within academic writing suggests a clarity and understanding of the world which analysts rarely have. Why then not use a form that reflects more clearly the plurality and partiality of perspectives within our research?

Laurel Richardson makes a similar argument from a more traditional sociological perspective. Her work on 'alternative representations' makes use of forms of (personalised) knowledge that are typically barred from the canon of valid social scientific research. Thus, she relates anecdotes, personal histories, and even dreams in an attempt to 'examine how knowledge claims are constituted in scientific writing, to write more engaged sociology, and to reach diverse audiences' (2002: 414). In doing this she situates herself, as author, within a particular time and space. This practice is in part politically motivated, but it is also, she suggests, a productive one for her research: 'Personal narration, reflexivity, and contextualization, I believe, are valuable tools' (ibid: 415). A number of geographers influenced by feminist and postcolonial theory have taken similar approaches (see, for example, Harding, 1991; Rose, 1997). Like Richardson, they do not pretend to be 'all-seeing and all-knowing researcher[s]' (Rose, 1997: 305), but instead recognise knowledge as situated and partial. mrs kinpaisby (2008) – a pseudonym for three academics engaged in participatory action research – takes these ideas further in attempting to address 'the continued disembodiment of academic voice' (p.298): they present their thoughts as a dialogue between the authors and invite the reader to 'eavesdrop' into their conversation.

We have taken this body of work to heart and seek within this chapter not to represent a static 'truth' about critical risk research but rather a set of particular and situated experiences active within a context increasingly conscious of 'risk'. In what follows we explore the ways in which the terminology of risk becomes woven throughout our empirical practices, exposing some of the normally hidden practices that characterise 'critical risk research'. In presenting our 'conversation' we also acknowledge that our 'method of writing' (Richardson, 2002) forms part of our knowledge and learning about risk.

The following discussion is presented as an interaction between the three authors of this piece: Sarah Davies (SRD), Brian Cook (BRC) and Katie Oven (KJO). The text shown is edited from a longer web-based conversation developed over a period of weeks in early 2010 and renewed in mid 2011, which began with the question: how does the concept of risk affect our research practices? From that point, the discussion followed a number of directions, eventually focusing on how 'risk' is articulated within our very different research contexts. Prior to the conversation, though, it is important that we briefly describe the domains in which our research and learning are situated.

SRD is a scholar of science, technology and society who studied natural science and worked as a science communicator before coming to research different practices of public engagement with science, including lay responses to emerging (and 'risky') technologies such as nanotechnology. Her work is situated in the UK's history of science communication and public engagement and its ostensible move from 'deficit to dialogue' (House of Lords, 2000). Her interests include the practice of public dialogue (such as the ways in which power is distributed within deliberative processes), the governance of new technologies, and the ways that laypeople construct responses to nanotechnology and other emerging technologies. The complexity and sophistication of lay talk is a key theme within her work, which therefore speaks to studies of public attitudes and opinion. In doing so it responds to a widespread tendency to imagine public perceptions of technology as calculative assessments of 'risks and benefits' (cf Wynne, 2001). Her interests in 'risk' therefore focus on the ways in which it gains meaning both within informal talk and in social technologies such as surveys and policy tools.

BRC is a social scientist who explores water management with an emphasis on risk, vulnerability and knowledge. His research spans the developing and developed world and integrates perspectives from traditional and non-certified experts. Running throughout his research are efforts to problematise 'prevailing' knowledge relative to alternate forms of knowledge and knowledge production. Questions of knowledge entrenchment and the power relations that privilege certain logics (Barry *et al.*, 2008) or framings (Miller, 2000) drive analyses that ask 'why and how certain forms of flood management persist?'. The research is grounded using the mutually constitutive (hybrid) relationship between knowledge, behaviour and context. Risk is a constant presence in his research, whether as a result of exposure to physical forces, in the perceptions of people 'living with risk', or in the wider concerns of people confronting socio-environmental and economic challenges.

For example, his work in Bangladesh uses interviews with power-holding experts with the aim of understanding the basis of the decision-making that underpins flood risk management. The research examines both straightforward questions about flood risk (i.e. is flooding a serious threat to Bangladesh or the people in Bangladesh?) as well as more complex questions about how experts interpret risk and, given that knowledge, implement risk management. Here, questions focus on the plurality of knowledge/interests, on the nature of resilience, and on the materiality of actants such as physical events and the structures used to 'control' flooding. In other research, he engages expatriates in Southern Portugal who are concerned with the risks associated with climate change. The research explores the transfer of knowledge of water regimes from temperate to semi-arid regions. In the UK, he works with Catchment Organisations (e.g. Tweed Forum, The River Dee Partnership) who are attempting to reconcile a desire for more natural flood mitigation (human-environment relationship) with economic, environmental and social change.

KJO is a geographer undertaking applied, interdisciplinary research in the natural hazards field. She has field experience in both developing and developed countries including Nepal, Taiwan, New Zealand and more recently the UK. KJO studied both physical and human geography before embarking on an interdisciplinary PhD investigating the vulnerability and resilience of rural communities to landslide and debris flow hazards in rural Nepal. For this she took a bottom-up, livelihoods-based approach to examine local understandings of, and responses to, mass movement hazards. She also explored institutional and expert understandings of landslide risk in Nepal and investigated how certain discourses around risk have become privileged and institutionalised. KJO's current research: 'Building Rural Resilience in Seismically Active Areas' investigates how to use scientific knowledge effectively alongside local knowledge with a view to building resilience to earthquake-related hazards.

The conversation

SRD: Getting a dialogue going when we are working in different continents is hard! But as a starting point I have put an opening question – how does the concept of risk affect our empirical research practices? – up on the discussion board. Please do post your thoughts on this, as well as other 'starting' questions which you think might be interesting.

KJO: To start, then: my PhD thesis, which examined the vulnerability of rural communities to landslide and debris flow hazards in the Nepal Himalaya draws, to some degree, on risk perception theory but mainly on the vulnerability and sustainable livelihoods literatures emanating from the development field. Before embarking on my empirical research, I spent time thinking through the different epistemological perspectives on risk. As a researcher interested in geophysical hazards, which pose a very real, and to a certain extent quantifiable, risk to society in rural Nepal, I immediately rejected the strong constructionist notion of risk. I did, however, gain a great deal from the socio-cultural literature on risk.

I entered the field with a particular vision of 'the problem'. Previous research had highlighted the concentration of landslide-induced fatalities in the Himalaya and, specifically, the high numbers of fatalities year on year in Nepal (Petley *et al.*, 2007). However, contrary to these ideas, when my fieldwork commenced it rapidly became clear that landslides were not a priority for rural householders in the case study communities. Residents are well aware that they live in a landslide prone area and have a good understanding of the physical causes and triggering mechanisms associated with landslide activity, but landslides are placed in the context of wider societal concerns including access to water and healthcare, employment opportunities and education. This made me question my research and, while landslides remained the focus of my study, I became far more critical and reflexive. The cultural perspective of the anthropologists Douglas and Wildavsky (1982) and Torry (1979) exposed to me the range of social, cultural and political factors that influence people's perceptions of, and responses to, hazard and risk.

I also drew parallels with Beck's 'Risk Society' thesis. A key factor driving the migration of households into landslide prone areas was the construction of a road in the valley bottom. People were drawn to the roadside by the economic opportunities associated with the highway and constructed houses at the bottom of steep, unstable slopes and adjacent to incoming stream channels prone to debris flow hazards. But while some risks are increased (risks associated with landslides), there are growing opportunities and reduced risks in other respects, for example, the opportunity to generate an income for a more secure livelihood and to access services. The construction of roads could be viewed as an example of technical innovation breeding risk as argued, in part, by Beck. Links can also be made here with the work of Gilbert White on floodplain encroachment (White, 1945). But, when it came to unpicking the complex decision-making process behind the occupation of landslide prone areas, I found the livelihood literature emanating from the development field to be particularly useful (Ellis, 2000; Rigg, 2006; Gough *et al.*, 2007).

BRC: For my research, I emphasise the knowledge that informs decision-making because it helps me to focus on the many factors that shape perception, knowledge and behaviour. Knowledge of risk is one factor, amongst countless, that contributes to decision-making and management. The people who inform my work engage with multiple risks in their daily (and work) lives: the risk of losing their job, of getting sick, of family troubles, and of exposure from hazards such as floods or earthquakes. By emphasising how they understand and relate to all of these risks, I attempt to understand the knowledge that informs/influences decision-making. It is an appreciation for competing knowledge – some of which is highly influenced by risk – that I analyse why and how people choose to behave. In this context, with technical flood management like dams or embankments comes knowledge that, simultaneously, affects how subsequent decision-making occurs. In visualising this relationship, I am thinking of networks and/or assemblages. In terms of the Risk Society thesis, I utilise the 'reflexive modernisation' argument, but as explained (possibly reinterpreted) by Adams (1995: 180). As he argues, it is important to appreciate that how we perceive and act upon risk is culturally filtered and continuously being modified. A second and related aspect of the discussion surrounding Risk Society that influences my research is the challenge to science and technology – and of scientists and engineers responsible for flood management – as arbiters of truth (here I am thinking of the expert-experience debate between Collins and Evans, 2002; Wynne, 2003 and Jasanoff, 2003). I have also used criticisms of Risk Society as inapplicable in the developing world as a way of showing how cultural or contextual differences (Haque, 1997; White, 1974) shape competing perceptions and knowledge of risk and hazard.

SRD: I wouldn't define myself as a risk researcher, but risk gets folded into my work in at least a couple of ways. I'm interested in the relationship between science and society, and in the ways that this is managed and understood. My research has looked at how public engagement and science communication events are carried out, and how publics construct meanings around emerging technologies. Nanotechnology is an interesting example as it's seen, by some at least, as extremely risky (Schefeule *et al.*, 2007): there are lots of concerns about the potential effects of engineered nanoparticles on the environment and on human health, for instance. But I've been interested more in how the idea of 'nanotechnology as risky' gets *used*, both within talk at the micro-level and more broadly in political discourse. So when I've looked at focus group transcripts of laypeople talking about nanotechnology, for example, it's clear that discussions of risk perform particular functions: as a kind of standard repertoire for dealing with unknown technological developments, or as a way of apportioning responsibility.

I have found that references to risk are fairly limited in lay talk. People tend to be more concerned about broader societal issues such as the unequal distribution of technological risks and benefits or their powerlessness to shape the development of technologies such as nano (Davies *et al.*, 2009).

At the macro-level, I'm even more interested in how risk is used. Much of my fascination with lay talk is a reaction to the way that risk can be used as a way of flattening the complexity of public responses to science and technology, rather than acknowledging and responding to it effectively. This is a political domain which, in the UK at least, continues to be difficult to negotiate and which – in the need for decision-making and soundbite policy stances – must continually fight a tendency to over-simplify; often, this over-simplification comes in the form of summing up scientific anxieties as 'uncertainty' and public ambivalence towards technology as misreadings of 'risk'. I have been influenced by the critiques of authors such as Brian Wynne and Steven Yearley who suggest that risk is often used as a fairly oppressive framework into which all public anxieties about science are categorised. When public concerns about science are assumed to simply be down to the fear of risk, other kinds of resistance – against science's corporate nature, its relative lack of accountability, or its ultimate aims and purposes, for instance – are effectively ignored.

BRC: KJO, I wonder whether there is not some contradiction between rejecting 'constructionist' interpretations of risk and accepting Douglas and Wildavsky's arguments? Is cultural risk theory not a good example of wanting to account for culturally-based constructions of risk?

KJO: It's the 'strong' constructionist notion of risk that I reject, not constructivism *per se*. I interpret strong constructionists as viewing hazards and risks strictly as social constructs. But to me, landslides present a very real and tangible risk to households and communities in Nepal. I do, however, recognise that risks are mediated through social, cultural, and political processes. I therefore see myself more as a weak constructionist. I mapped the landslide and debris flow hazards in the Valley where I undertook my research, providing an outside scientist's view of 'the problem'. I then conducted household surveys, interviews and participatory mapping exercises to gain the perspective of the people deemed to be 'at risk'. My assumption was that taking a solely top-down or bottom-up approach can result in a one-sided account of the problem and may lead to inappropriate solutions (I'm thinking here of Fairheads and Leach's work (1995) challenging environmental narratives in Africa). In the context of landslides in Nepal, zoning the road corridor and making it illegal to settle in landslide prone areas may have a negative effect on the households occupying these locations. But, ignoring the landslide problem altogether would leave households and communities susceptible to potentially fatal events.

BRC: In reading this, it reminds me of my struggles with 'realist vs. relativist' discussions. I don't see any need to presume they cannot be compatible. I find that a 'one or the other' view does not reflect what I study and we are often confronted with situations in which a realist perspective is needed *and* where we have to appreciate the socially constructed nature of the knowledge-power-world.

To get back to risk, I thought it might be helpful to look the word up on the online Oxford Dictionary. Here's what it says:

Noun. 1: a situation involving exposure to danger. 2: the possibility that something unpleasant will happen. 3: a person or thing causing a risk or regarded in relation to risk: a fire risk.

Verb. 1: expose to danger or loss. 2: act in such a way as to incur the risk of. 3: incur risk by engaging in (an action).

In each of these examples, risk is not a way of thinking or a worldview. This makes me consider whether the 'risk society' is really not something more like the 'knowledge society' – i.e. it is the extension of our knowledge that leads to changes in behaviour. Growth in knowledge therefore leads to reconceptualisations of society or the environment, for instance. I've been trying to think about this with regard to the idea that we might use risk to understand the outcomes of our work rather than seeing it as an instigator of it. While I explore risky situations, risk is not the theoretical lens that shapes my view of the world; rather, it is part of the assemblage that instigates change or evolution at the human-environment interface.

Re-reading that, I'm not happy. Risk is part of how we understand, it is a force that shapes knowledge and behaviour, it is a lens (or shaper of lenses), and it is a reflexive agent. I suppose what I'm getting at is what we call 'risk' is actually many things, in many contexts, and subject to many interpretations.

SRD: I think this is a key point (and one which speaks to my interests in the different ways risk can be constructed and used). To build on it: I'm really interested in the different storylines of our research, and the varying tractions 'risk' has on our data. How are concepts of risk used in different ways?

KJO: I drew heavily on the work of Frank Ellis and Jonathan Rigg, who have both written on rural livelihoods in developing countries. Ellis' work is concerned with household decision-making and livelihood diversification including migration and the familiar 'push/pull' debate. Similarly, Rigg's research focuses on everyday geographies of the Global South. For Ellis and Rigg, risk wasn't/isn't their starting point. Instead, their research is concerned with understanding the complexity of the everyday and risk is part of this. From my perspective, Ellis' and Rigg's work falls into the category of risk theory too.

SRD: I think your uncertainty about what, exactly, 'counts' as risk research is interesting. It seems to me that risk is one of those words which start to lose their meaning the more that you think about them – like the games children play by repeating a single word over and over again until it starts to sound completely alien. If I were to do a word association for 'risk', I suspect I would come up with things like 'reflexive modernity', 'uncertainty', 'hazard', 'danger', and 'risk perception', but I'm also influenced by Deborah Lupton's (1999) work, which neatly (possibly too neatly) subdivides risk theory – at least in terms of socio-cultural rather than realist approaches to it – into risk society, cultural and Foucauldian approaches. But really what we seem to be saying is that all sorts of different kinds of work – from development geography to actor network theory – is actually useful to us in studying 'risky' places and situations. We are all, I guess, working in more or less interdisciplinary contexts, and perhaps one hallmark of such contexts is that there is a more established sense of bricolage in the literatures we draw on (Donaldson *et al.*, 2010).

One thing that also seems to be connected to risk in all of our work is the notion of expertise, and who is understood as an expert in any particular field or topic.

BRC: For my work in Bangladesh, people were often in favour of science and scientific expertise, appearing to have faith in its ability to deliver a better standard of living. This makes me wonder about the claim that 'Risk Society' risks are somehow different from modern or industrial society hazards. Industrial risks and pollution in the developed world were also, in their time, very damaging, and prompted a reconceptualisation of what was acceptable or manageable. It was only over time that people were able to understand and mitigate impacts. While I am not suggesting that technology will solve all the world's ills, it is a little premature to suggest that we cannot or will not be able to do the same for BSE, climate change or pesticides. This connects to the notion of risk transfer or displacement and, to some degree, to Adam's (1995) thermostat argument. This suggests that risk is not (or is rarely?) eliminated, but is instead changed through the process of management. It might even be argued that it is the act of management that prompts, or prompts recognition of, new or previously hidden risks. Admittedly, we do not know what management will uncover or initiate, and I suppose that is the point. It is the uncertainty in terms of risk mitigation impacts that, I suppose, becomes important. I spoke with flood managers in Bangladesh who were frustrated with the 'silo' or disciplinary nature of flood management because it 'lost track' of the risk as it floated between agencies or ministries.

They explained how risk 'travelled' through the Bangladeshi population (rising oil prices, leading to rising fertiliser prices, leading to rising rice and cooking oil prices, leading to malnutrition, social unrest, poor agricultural practices, exploitation of people by those with money/power), which suggests to me that, like flooding in Bangladesh, risk is beyond control.

I would suggest that the developed world's emphasis on 'transferred' or 'previously hidden' risks is a psychological response to our ability to – largely – control low and medium scale 'natural' hazards that remain common in much of the world. I suspect that our minds have been wired with a disposition that is continually searching for risk to ensure our safety; I would suggest that it will find risk in any situation, leading those in relative safety to find threat in less evident contexts. I say this not to undermine the threat of radiation or pesticide use, but to show that wealth has afforded us the opportunity to recognise other risks. These risks exist in the developing world as well, they are simply overshadowed by other threats – among them subsistence and everyday life. I suppose the key distinction, and one that Beck and others who discuss the Risk Society make quite clear, is that expertise becomes key to interpreting, but also bound up with producing, these 'new' forms of risk. While flooding is evident, causing immediate and obvious impacts, BSE, climate change, and the flushing of hormones into waterways require a high level of expertise to recognise, measure, or understand.

Coming back to scientific expertise, the premise that there is a popular resistance or criticism of technology – which is how I read Beck – does not reflect my experiences in Bangladesh. On the contrary, I witnessed people with great faith in technological solutions to their problems; whether these assessments included appreciation for the potentially detrimental consequences associated with the technologies, I cannot say. There is a reflexivity amongst the scientists and engineers with whom I worked, which amounted to their recognising the need to engage with the public (or at least recognise the need to say this). This is one of those instances where my presence as a social scientist interested in environmental management and engineering may have influenced what was said; despite this, the engineers in particular were adamant that they had learned from past criticisms and now valued public scrutiny and participation. While their meaning of participation was sometimes more akin to informing the public about what the experts thought, there were examples of genuine collaboration between decision makers and the public.

KJO: My experience in Nepal was much the same. I found the division between 'lay and expert' and 'local and outside' knowledge in the context of landslide risk management to be somewhat artificial. Whilst the geologists and engineers argued that their expertise was often ignored in favour of fashionable participatory approaches, they recognised people's decisions to occupy landslide prone areas as largely rational. Overall, I believe they understood the perspectives, capacities and priorities of rural people. The majority of scientists I interviewed recognised the knowledge that lay people have of the physical environment. This knowledge is derived from their everyday, lived experience. For example, communities are known to have a very good understanding of monsoon triggered landslides that are a regular occurrence across the hill villages of rural Nepal. There are, however, significant knowledge gaps around rare, high magnitude events such as earthquakes that have not been directly experienced. Similarly, the migration of people to new areas can leave them exposed to new, unfamiliar hazards such as debris flows. The scientists interviewed certainly believe they have a role to play in raising awareness of these unknown risks for example, by advising on the most appropriate alignment for a new road.

Long and van der Ploeg (1994) note that knowledge is based on both indigenous experience and introduced techniques, as outside technology and ideas are reworked. It is perhaps unsurprising therefore that householders interviewed felt that engineering was the key to managing large slope failures deemed to be beyond the control of the community. Frequently cited examples include the construction of retaining walls and check-dams which have been used to protect vulnerable sections of the highway from mass movement events. In general, local people felt that such engineering solutions should be used to stabilise slopes and protect against debris flows in the case study settlements.

SRD: There seem to be some similarities between the contexts in which you both work – or perhaps the framings you use? – and the risks viewed as important, although the approaches you take are slightly different. In contrast, the research I've done has largely been around emerging technologies and has been done entirely in the US and UK. While there has been a similar participatory turn – with science policy increasingly using methods of public deliberation and engagement in order to structure the development of research – concepts of risk, and of who is expert in thinking about those risks, are a little different, not least because the riskiness of nanotechnology, for instance, is currently largely invisible to the publics who may eventually be affected by it (Schefeule *et al.*, 2007). Public awareness is so low – understandably, given that many of these technologies are not going to be available for years or decades, if ever – that there is no 'local knowledge' in the way that there might be around earthquakes or floods, for instance. This is not to say that people don't have views about how nanotechnology or other kinds of technology should be developed. But concern about hazards to human health and the environment tend to come from scientists and other technical experts, while laypeople are more concerned about effects on society, relationships, and sense of self (Davies *et al.*, 2009). I'm not sure whether anxiety regarding the unequal spread of useful technologies can be construed as anxiety about risk, for instance, or whether exploring the notion of meddling with nature classes as critical risk research.

Because of this, I continue to be somewhat wary of the ways in which 'risk' is used. While much risk research is clearly extremely sophisticated in how it thinks about who has expertise around a particular hazard, in other contexts discussion of risk can simply lead to an automatic deferral to the authority of scientific experts. Brian Wynne has argued this particularly strongly, suggesting that orienting the management of science in society around 'risk and ethics' leads to an emphasis on technical risk assessments and the blackboxing of wider public concerns (Wynne, 2001).

BRC: I am struck by the comment about risk as a way to simplify and impose a structure on knowledge, particularly for the purpose of governing. Risk, in this sense, can be understood as a tool to gain leverage over people. I've certainly run into this. In many instances, governance is steeped in discussions of risk as a way of saying to the public: 'you may not like what we are proposing/doing, but you should accept it because the world is a risky/threatening place'. Whether or not what is being done has to do with risk directly, or is part of wider government objectives is the point I am trying to get at. For example, I was present when a proposition for a ring road/embankment to encircle Dhaka was being discussed; the primary arguments both *for and against* this massive infrastructure was that the displacement of water would also displace risks from the relatively affluent inside to the impoverished outside. While speaking with the critics of the proposal afterwards, it was clear that they had a wider dislike or scepticism towards technical management, but that the risk argument resonated with the public and was therefore used to undermine strategies they characterised as 'techno-centric', 'led by engineering consultants' and 'a poor use of limited resources'. Risk provided the language for the debate, but it was other issues that provided the impetus.

SRD: Again, you see risk featuring as a rhetoric – a discursive device – in order to bring about a particular conclusion. It interests me that the people you spoke to were overtly conscious of this use of risk, as well. In focus groups on nanotechnology 'risk talk' at times seems to function as a semi-automatic response, a well-rehearsed way of discussing uncertainty, but there is little overt exploration of risk as a strategy. Perhaps the stakes are higher when there is an immediate decision to be made and the context is very clearly political.

To change the subject, somewhat, I feel that we should reflect, even if it's just briefly, on what experimenting with this format has been like. What different kinds of things have emerged from working and writing like this?

BRC: In thinking about this chapter, I am both very happy to have experienced the discussions but also a little hesitant that others will read it. There is something protective about the conformity of academic writing, hiding the blemishes and making sure we don't step too far out of line. That said, it is useful to 'hear' the thoughts and witness the thinking that goes into scholarly opinion and research. I've also found it very challenging to come back to a discussion which can later appear simplistic or misinformed given how my thinking has evolved. I am pleased to see the hesitancy that seems to challenge the claims and arguments. It is easy to forget that academic writing is a sanitised product that is produced over months (or longer) and has had numerous revisions and people to provide comments. I suspect that there is a reason that this type of 'conversation' in an academic context is not a common format; that said, its differences provide insight into the dominant format. Perhaps because of the nature of risk or because of the material elements of critical risk research, our discussion suggests that risk is something that instigates, contextualises and influences my research at every stage.

KJO: I thought this was a very interesting exercise. As noted by Adams and Megaw (1997) '[t]oo often, our accounts of what we have done draw on selective memory and a desire to conform and convince a critical First World audience' (p.216). Working in this format provided me with an opportunity to present an honest and reflexive account of my research experience and engage with others in a discussion that brings what mrs kinpaisby (2008) describes as the 'ordinary voice' [of SRD, BRC and KJO] into academic debate. Our candour, I think, has helped us avoid the 'false neutrality and universality of so much academic knowledge' (Rose, 1997: 306). However, in taking up this reflexive position, I have, at times, wondered if my lack of engagement with 'traditional' risk theory reflects my own ignorance of it. While I consider this an important admission, I remain mindful of the fact that my research was a study of landslide vulnerability that emphasised the positions and perspectives of vulnerable people. The theory, then, that most influenced my empirical research practice comes from anthropologists and geographers in the hazards and development field.

SRD: I always find this kind of interaction very freeing – there's much more scope to discuss ideas which are messy or inchoate, or which go against dominant opinion in a discipline. I'm also fascinated by *practices* of research, so it's interesting to hear the nitty gritty of other people's work (Katie changing her mind about the most important 'problem' her research participants had to deal with, for example). And I've definitely enjoyed being liberated from writing in the stiff, unnatural style that much academic text has (and which is, I think, more about the production of a sense of authority than portraying the research process). However, I've also been very aware that even this kind of text is not a straightforward portrayal of the 'reality' of research. Our 'conversation' was stage-managed, and cutting it up and sticking it back together to turn it into this book chapter was a fairly big job. The final result certainly references one way of doing research, but it may also be just as artificial as more formal presentations.

Conclusion

To conclude by again picking up our shared and abstracted authorial voice: this conversation has emphasised that there are multiple interpretations and assumptions involved in critical risk research. While this is not an original finding, the need to describe and clarify portions of a transcribed conversation has allowed us to critically engage with assumptions that would normally have been 'flattened' by the presumption of a shared understanding of risk. Conversations flow with the freedom of a certain lack of accountability; this allows new ideas to flourish but also risks allowing poor ideas to take root. Differences are less obvious, particularly with regard to popular terms and concepts; this conversation – recorded and then mulled over over a period of months – brought such differences into focus. More than anything we have been forced to revisit assumptions that had come to underpin different uses of the same term, 'risk'. Inconsistencies that might have otherwise gone unquestioned were exposed through the act of description and clarification.

Our research, despite utilising a common concept, applies differing assumptions, definitions, literatures and objectives. A first reaction might be to call for clarity so that individuals can communicate more effectively. But we believe the benefits of plurality are critical if risk is to continue to positively influence thought and research. Rather than call for uniformity we would argue that our perspectives have been both broadened and strengthened by exposure to alternate voices.

References

Adams, J. (1995) *Risk*. London, Routledge.

Adams, W. M. and Megaw, C.C. (1997) Researchers and the rural poor: asking questions in the Third World. *Journal of Geography in Higher Education*, **21**(2), 215–229.

Bankoff, G., Frerks, G. and Hilhorst, D. (Eds.) (2004) *Mapping Vulnerability: Disasters, Development and People*, Earthscan, London.

Beck, U. (1992) *Risk Society: Towards a New Modernity*, Sage, London.

Collins, H.M. and Evans, R. (2002) The Third Wave of Science Studies: Studies of Expertise and Experience. *Social Studies of Science*, **32**, 235–296.

Cutter, S. L. (1996) Vulnerability to environmental hazards. *Progress in Human Geography*, **20**(4), 529–539.

Cutter, S.L., Mitchell, J. T. and Scott, M.S. (2000) Revealing the vulnerability of people and places: a case study of Georgetown County, South Carolina. *Annals of the Association of American Geographers*, **90**(4), 713–737.

Davies, S.R., Kearnes, M.B., and Macnaghten, P.M. (2009) All things weird and scary: Nanotechnology, theology and cultural resources. *Culture and Religion: An Interdisciplinary Journal* **10**(2), 201–220.

Donaldson, A., Ward, N. and Bradley, S. (2010) Mess among disciplines: interdisciplinarity in environmental research. *Environment and Planning A* **42**, 1521–1536.

Douglas, M. and Wildavsky, A. (1982) *Risk and Culture: An Essay on the Selection of Technological and Environmental Dangers*, University of California Press, Berkeley, California.

Ellis, F. (2000) *Rural Livelihoods and Diversity in Developing Countries*, Oxford University Press, Oxford.

Fairhead, J. and Leach, M. (1995) False forest history, complicit social analysis: Rethinking some West African environmental narratives. *World Development* **23**(6), 1023–1035.

Gough, I., McGregor, J.A. and Camfield, L. (2007) Theorising wellbeing in international development, in *Wellbeing in Developing Countries From Theory to Research* (eds I. Gough and J. A. McGregor), Cambridge University Press, Cambridge, pp. 3–43.

Haque, C.E. (1997) *Hazards in a Fickle Environment: Bangladesh*. Springer, Boston.

Harding, S. (1991) *Whose Science? Whose Knowledge?* Cornell University Press, New York.

House of Lords (2000) Third Report: Science and Society, The Stationery Office, Parliament, London.

Jasanoff, S. (2003) Breaking the Waves in Science Studies: Comment on H.M. Collins and Robert Evans, 'The Third Wave of Science Studies'. *Social Studies of Science* **33**, 389–400.

Latour, B, and Woolgar, S. (1986) *Laboratory Life: The Construction of Scientific Facts*, Princeton University Press, Princeton.

Law, J. (2004) *After Method: Mess in Social Science Research*, Routledge, Abingdon.

Leach, M. and Scoones, I. (2005) Science and citizenship in a global context, in *Science and Citizens. Globalization and the Challenge of Engagement* (eds M. Leach, I. Scoones and B. Wynne), Zed Books, London, pp. 15–38.

Long, N. and van der Ploeg, J.D. (1994) Heterogeneity, actor and structure: towards a reconstitution of the concept of structure in *Rethinking Social Development: Theory, Research and Practice* (ed D. Booth), Longman Scientific and Technical, Harlow, Essex.

Lupton, D. (1999) *Risk*. Routledge, London.

Macnaghten, P. and Urry, J. (1998) *Contested Natures*, Sage, London.

McDowell, L. (1992) Doing gender: feminism, feminists and research methods in human geography. *Transactions of the Institute of British Geographers*, **17**(4), 399–416.

mrs kinpaisby (2008) Taking stock of participatory geographies: envisioning the communiversity. *Transactions of the Institute of British Geographers*, **33**(3), 292–299.

Mulkay, M. (1991) Textual fragments on science, social science and literature, in *Sociology of Science: A Sociological Pilgrimage* (ed M. Mulkay), Open University Press, Buckingham, pp. 21–36.

Petley, D.N., Hearn, G.J., Hart, A., Rosser, N.J., Dunning, S.A., Oven, K.J. and Mitchell, W.A. (2007) Trends in landslide occurrence in Nepal. *Natural Hazards*, **43**(1), 23–44.

Richardson, L. (2002) Writing sociology. *Cultural Studies <=> Critical Methodologies*, **2**(3), 414–422.

Rigg, J.D. (2006) Land, farming, livelihoods, and poverty: rethinking the links in the Rural South. *World Development*, **34**(1), 180–202.

Rose, G. (1997) Situating knowledges: positionality, reflexivities and other tactics. *Progress in Human Geography*, **21**(3), 305–320.

Scheufelel, D.A., Corley, E.A., Dunwoody, S., Shih, T-J., Hillback, E., and Guston, D.H. (2007) Scientists worry about some risks more than the public. *Nature Nanotechnology* **2**(12), 732–734.

Torry, W. (1979) Hazards, hazes and holes: A critique of *the Environment as Hazard* and general reflections on disaster research. *The Canadian Geographer*, **23**(4), 368–383.

Tulloch, J. and Lupton, D. (2002) Consuming risk, consuming science: The case of GM foods. *Journal of Consumer Culture*, **2**(3), 363–383.

White, G.F. (1945) *Human Adjustment to Floods: A Geographical Approach to the Flood Problem in the United States*, Chicago University Press, Chicago.

White, G.F. (1974) *Natural Hazards, Local, National, Global*, Oxford University Press, New York.

Wisner, B., Blaikie, P., Cannon, T. and Davies, I. (2004) *At Risk: Natural Hazards, People's Vulnerability and Disasters*, Routledge, London.

Wynne, B. (2003) Seasick on the Third Wave? Subverting the Hegemony of Propositionalism: Response to Collins & Evans (2002). *Social Studies of Science*, **33**, 401–417.

Wynne, B. (1996) May the sheep safely graze? A reflexive view of the expert-lay knowledge divide, in *Risk, Environment and Modernity: Towards a New Ecology* (eds S. Lash, B. Szerszynski, and B. Wynne), Sage, London, pp. 44–83.

Wynne, Brian (2001) Creating public alienation: expert cultures of risk and ethics on GMOs. *Science as Culture*, **10**(4), 445–481.

PART 2
Politics in Risk Research

CHAPTER 5

Finding the Right Balance: Interacting Security and Business Concerns at Geneva International Airport[1]

Francisco R. Klauser[1] *& Jean Ruegg*[2]

[1]Institut de Géographie, Université de Neuchâtel, Neuchâtel, Switzerland

[2]Institut de politiques territoriales et d'environnement humain, Université de Lausanne, Lausanne, Switzerland

Introduction

'Risks are man-made hybrids', Ulrich Beck writes; 'they include and combine politics, ethics, mathematics, mass media, technologies, cultural definitions and precepts' (Beck, 1998: 11). In other words, risks are framed in specific ways, by specific people, needs and interests. Researching the causes, modalities and effects of particular risks, thus, always requires critical appreciation of how, why and by whom the problems at hand are defined and how these definitions then legitimise particular interventions. Yet whilst the socio-political constructions and exploitations of risk have been acknowledged on various conceptually and empirically informed grounds, there is to date a pressing need to better understand the precise ways in which the various interests in, and practices of, risk management merge (in consensus and conflict) within a particular milieu, and the ramifications this has. What is needed is a micro approach that allows an understanding of how exactly risk management, as the outcome of

[1] This article expands upon empirical results from the research project 'Vidéosurveillance et risques dans l'espace à usage public: représentations des risques, régulation sociale et liberté de mouvement', financed by the Swiss National Science Foundation (Ruegg *et al.*, 2006). This research has been conducted by Jean Ruegg, Valérie November, Francisco Klauser and Alexandra Felder (Social Sciences) and by Alexandre Flückiger, Laurence Greco and Laurent Pierroz (legal research group).

Critical Risk Research: Practices, Politics and Ethics, First Edition.
Edited by Matthew Kearnes, Francisco Klauser and Stuart Lane.

complex interactions and negotiations, permeates and shapes particular places and points in time.

Drawing upon the empirically based study of the securitisation of Geneva International Airport, our chapter sets out to address precisely this issue. More specifically, we seek to explore some of the concerns, interests and risk perceptions underpinning the policing of the public premises at Geneva Airport (i.e. the arrival and departure halls and the airport railway station). In this endeavour, particular emphasis will be placed on the complex relationships between the security concerns and the business interests associated with the public airport sections, and on how these relationships are shaping the everyday security operations of the airport police (*Police de la Sécurité Internationale*[2]).

In what follows, thus, we have chosen to focus on a particular set of interacting concerns in airport security. However, we are well aware that the imperatives of airport security are in reality much more complex and cannot be explained comprehensively by such an intentionally limited approach. We are not implying here that airport security is shaped exclusively by its multiple connections and interactions with business interests. We merely aspire to provide a symptomatic, if necessarily limited, illustration of the alliances, tensions and dilemmas in contemporary security governance and hope thereby to show that the modalities and effects of risk management, and more specifically of security governance, are inevitably both ambivalent and ridden with contradictions. The values and effects of security governance are inherently complex and diverse, and depend on the everyday micro negotiations through which specific security measures are pursued and co-constructed. Finding the right balance in security governance is not so much a question of universal principles than of hazy everyday negotiations and micro adjustments between different actors, interests and issues. In this everyday balancing and recalibrating exercise of security governance, critical risk research can, and must, play an important role.

Methodology

Our investigation draws upon empirical insights provided by a two-year research project funded by the Swiss National Science Foundation (Ruegg *et al.*, 2006). Based on several case studies, and bringing together social scientists and legal specialists, the project examined the multiple factors

---

[2] Amongst many other duties, the airport police, an entity of the cantonal police at Geneva, are commissioned with passport and border controls at the airport, as well as with the 'the securitization of the airport installations, official buildings, tarmac, runways, airplanes on the ground and of the airport territory' (Etat de Genève, 2001: Art 2).

contributing to the way contemporary security governance functions and the impact it has.

As regards the case study of Geneva International Airport more specifically, 11 open-ended, qualitative interviews with the key stakeholders in the planning, installation, use and development of the airport security system were conducted. Interviewees included the current and former heads of airport police, the head of the airport police control room and two police CCTV operators, as well as representatives from the airport's technical and legal services, the CCTV system suppliers and local political authorities. In addition, everyday security practices at the airport were studied during one week of observational research in the airport police control room. This methodological approach will not be explored in this chapter, however.

In order to give a strong focus to this chapter, the analysis that follows will be limited to those parts of the interviews that are relevant to the issues associated with the interactions, coalitions and tensions unfolding between the public and private actors involved in the securitisation of the publicly accessible premises of Geneva International Airport.

Positioning

The research approach pursued here focuses on the micro level, locating the various security and business concerns in the airport in the context of a specific range of projects and decisions. Yet our aim is not only to provide isolated insights into the micro-politics of security and surveillance at Geneva International Airport, but also to re-institute this question as part of a broader set of issues: the complex relationships between security governance and business interests.

In recent years, this research problem has sparked a sophisticated body of literature. In view of the analysis that follows, it is worth mentioning two fields of research in particular. On the one hand, a number of scholars have pointed towards the critical influence of surveillance and securitisation strategies with regard to the privatisation and commercialisation of particular places. Such places include city squares (Franzen, 2001; Coleman, 2004), parks and transport hubs (Töpfer, 2005), alongside shopping malls and other publicly used yet privately owned places of consumption and leisure (Sorkin, 1992). From this perspective, important insights have been gained into the wider socio-spatial effects of security governance on the urban 'spheres of the everyday' (Klauser, 2010) and, more specifically, into a range of related issues in terms of social inclusion and exclusion (Koskela, 2000).

On the other hand, a growing body of research has in recent years focused on the logics and effects of the 'security business' itself, hence exploring the critical role of private authority and expertise in contemporary security governance (Lyon, 2004; Stevens, 2004). Yet despite the important insights provided regarding the role of private interests and responsibilities in current security matters, there is to date little empirical evidence of how exactly the public-private alliances underpinning particular security systems are working on an everyday basis.

The analysis that follows aims to link these two bodies of literature. By doing so, our chapter also complements two earlier papers, in which some of the empirical materials explored here have been approached from other conceptual and thematic perspectives (Klauser, Ruegg and November, 2008; Klauser, 2009).

Content

Dealing with the interacting security and business concerns associated with the public premises at Geneva International Airport, our study shows that security practices result from processes involving a range of actors, guided by converging and diverging goals, acting from mutually enhanced yet also potentially conflicting positions, and driven by shared benefits, whilst also pursuing their own specific agendas and projects. This study is divided into three main parts.

Firstly, the chapter sets out to investigate the coalition of interests between the airport management and the airport police, which underpins the securitisation of Geneva International Airport. In this perspective, security operations will be studied as a means to serve two main airport functionalities: the airport's position as a national entrance gate of critical symbolic and economic importance (1) and the airport's meaning as a zone of increased commercial activities in its own right (2).

Secondly, the chapter looks in empirical detail into the delicate and constantly renegotiated balance between the security concerns of the airport police and the business interests of the airport management. Based on a discussion of a series of micro illustrations, this analysis indicates the tensions between the imperative of achieving risk-free places on the one hand and the need to serve business interests on the other.

Thirdly, based on interviews conducted with the company supplying the airport CCTV system, the chapter focuses on the role of private technology companies in the 'making' of airport security. Here the chapter moves beyond the study of the specific interactions between the airport police and the airport management, to look in more detail into the role played by external actors and interests in the securitisation of Geneva International Airport. Although it will not be possible here to give an exhaustive interpretation of *all* the actors and needs in airport security, this investigation identifies the need to approach the modalities and effects of contemporary security governance by studying not only the converging and diverging concerns of locally anchored actors, but also the role of external forms of expertise, interests and authority.

Public-private coalitions of interests in airport security

Airport functionalities

The primary function and *raison d'être* of airports is mobility. Airports are to be understood as critical infrastructural and mobility hubs of increased national and economic importance, enabling and processing national and international flows of people and cargo. Consequently, most airport studies have tended to apprehend 'the airport' as a 'space of flows' (Castells, 1996) and 'perpetual transit' (Fuller, 2003). From a security perspective, a growing literature has sought to critically reflect upon the far-reaching implications of the extended and redesigned filtering and screening of international mobilities through airports (Lyon, 2003; Adey, 2004; Salter, 2008).

The Geneva case study confirms the 'mobility aspect' of airports. In 2009, Geneva International Airport was linked with over 130 destinations by around 50 scheduled airlines, transporting 11.3m arriving and departing passengers (Aéroport International de Genève, 2009: 18). As one of the police CCTV operators at the airport stated in our interview:

> It's true, we're a portal here. Everybody coming from afar passes through the airport. The airport is a real platform (Airport CCTV operator, translated by the authors).

However, the Geneva case study also highlights the fact that airports are not only detached worlds of transit of their own logic, but can also – more banally – become common places for local residents, tourists and other passers-by (Klauser, Ruegg and November, 2008). As a relatively small and easily reachable airport near the city centre, Geneva International Airport houses many shops with attractive opening hours, aimed at the general public. There are also regular performances and events (such as flea markets, fashion shows, exhibitions etc.), taking place mostly (but not exclusively) in the airport's integrated railway station, in order to enhance the 'airport experience' (Aéroport International de Genève, 2008: 30). This additionally attracts potential customers for the shops and restaurants in the airport's public zones. Furthermore, given the airport's location next to both a major exhibition centre and a music venue, some of its premises also accommodate flows of concert goers and other passers-by. Geneva Airport must hence be considered as a functionally diverse space, providing not only a range of services for passengers but also attracting people without any travel intentions, from local residents to tourists and concert goers to exhibition visitors. The airport's recent marketing initiative, 'GVA+', bears striking testimony to its multi-functional vocation:

> More airy. More space. More time. More choice. More flavour: these were some of the communication campaign slogans accompanying the implementation of AIG's [Aéroport International de Genève] masterplan. The GVA+ 2008-2010 programme's philosophy is to put the passenger and visitor at the centre of the airport (Aéroport International de Genève, 2008: 30).
>
> In its first phase GVA+ involved: press adverts highlighting client benefits such as more space and choice; and a new website, www.gvaplus.ch, covering changes, shop openings and events. While aimed at passengers, these initiatives are also designed to expand the clientele of AIG's commercial centre [by attracting travellers' companions, local residents and personnel from the airport and from its environment[3]] (Aéroport International de Genève, 2008: 42).

In the following, we shall briefly investigate how the two main functionalities of Geneva International Airport – as a national entrance gate and as a commercialised destination in its own right – resonate with the security operations of the airport police.

Airport CCTV as a means to fight against petty criminality

From different perspectives, the interviews conducted underscore the coalition of interests between the airport police and the airport management, who between them require the airport to be both a safe national entrance gate and a commercially appealing space for shopping and consumption. The following account, relating to the original driving forces in the installation of the first 13 police cameras in the check-in and arrival halls of the airport in 1996, provides an exemplary illustration of this:

> At the time we installed CCTV in the airport's public premises, a series of petty crime issues were apparent. The police said that they neither had the means to increase human presence nor to install further technical equipment. Since the airport management was afraid that Geneva Airport would become a context for all sorts of petty theft, negotiations had to set in: 'If we pay for the equipment, will you be ready to put somebody behind the video screens?' Yet I'd think that this was much more a matter of dialogue than an application of strict rules. Eventually, it was done like this and everybody was happy (Senior member of technical services at Geneva Airport, translated by the authors).

---

[3] This part is missing in the official English translation of the Airport's Annual Report. It has been translated by the authors from the French version of the report.

The above account reveals that CCTV at Geneva International Airport was born from a locally-anchored coalition of interests, joining together the police and the airport management. Rather than being imposed externally, CCTV was developed, framed and 'built up' through a series of internal exchanges and bottom-up planning processes as a necessary policing measure in the fight against the rising problem of petty crime. This is important to note for the purpose of this chapter, as it provides a good starting point from which to consider the joint and interactive approach to airport security, bringing together different stakeholders, interests and needs.

The quote above further points out that the risks of, and intervention against, petty crime, were defined and tackled in informal dialogue, rather than through the application of strict rules. Eventually, we see that from this locally-anchored dialogue, a hybrid situation of shared competences and authority arose: the police would operate CCTV, whilst the airport would cover all relevant material and installation costs. This situation has remained unchanged throughout the many developments and adaptations of the system, despite the fact that the exact modalities of collaboration have never been the object of any detailed formalised agreement (Ruegg *et al.*, 2006).

In sum, this initial discussion testifies to the merging concerns in airport security, and to the resulting mutually intertwined, informally negotiated positions held by the police and the airport management. For the police, in the fight against petty criminality, both the security of the airport and its reputation as a highly symbolic national entrance gate are at stake, whilst for the airport management, these issues also convey a business dimension. Both the police and the airport management converge in their endeavours to create a secure and attractive airport environment.

If we are to understand how exactly this coalition of interests in airport security permeates the everyday regulation of the airport, we must also investigate the actual policing measures and practices in the airport. At this point, however, a comprehensive overview of the whole panoply of airport security operations would be beyond the immediate scope of this chapter (for a larger discussion, see Ruegg *et al.*, 2006). Thus here we simply propose to discuss three aspects of airport policing, which are of particular relevance regarding the airport's position as both a safe national entrance gate *and* as a commercially attractive shopping zone.

A crime-free airport

The use of CCTV in the fight against petty criminality provides an initial framework within which to consider how everyday airport policing is shaped by converging goals and shared benefits. To investigate this further, we will first look into the influence of petty criminality on the planning and design of the CCTV system, before moving on to discuss the use of cameras for crime detection.

Regarding the planning and design of the airport CCTV system, our interviewees repeatedly emphasised the strong link between the spatial distribution of petty criminality and the location of the cameras. For example, no cameras were installed in the transit zone of the airport, given the low level of petty criminality in this area. In the publicly accessible zones of the airport, however, cameras were located strategically to cover exactly those places where luggage thefts and pickpockets were observed to occur most often.

> Cameras were not just installed anywhere, anyhow, but depending on those locations known as crime sites, where, obviously, we had recurrently encountered bag-snatchings and pick-pocketing. Furthermore, to identify these sites, a study was conducted by the airport management and the police. On this basis, we were able to determine where cameras should be placed (Former head of airport police and initiator of airport CCTV, translated by the authors).

Of course, the system's focus on petty criminality also finds expression in the everyday *real-time* camera operations of the police (screening the airport in search of thieves or suspect behaviour). In interview, CCTV operators described real-time crime detection as the most challenging, yet also most gratifying, activity, since they could rely directly on their experience and gradually accumulated understanding of where, when and by whom thefts are committed.

> Whenever we are behind a camera, we are searching for criminals. We're hunters with cameras. That's the aim (Airport CCTV operator I, translated by the authors).

Furthermore, it appears from our interviews that, over time, the operators' practical expertise in the fight against petty criminality not only shaped everyday camera operations, but also constituted one of the key factors in the various developments and adaptations of the system itself. Indeed, most of the gradual changes to the system (additional and repositioned cameras, or technical adaptations) were said to be driven by needs emerging directly from the everyday surveillance practices of the police.

This argument can be further developed by looking at the positioning of the cameras whilst unattended (at night, or when operators are engaged in other activities). Again, the system's focus on petty criminality, and the operators' experience in this field, appear to be of critical importance.

> We know that there are strategic points where more luggage robberies will occur than elsewhere. In the evenings, I focus the cameras especially on these points.

> Afterwards, if we have to visualize the images, I know that the cameras were already watching these points. We also try to have wide camera angles, in order to see the maximum possible (Airport CCTV operator II, translated by the authors).

In sum, the meaning of CCTV in the fight against petty criminality has not only shaped the original design and gradual adaptations of the system, but has also channelled the everyday policing of the airport. Thus the history of airport CCTV, and hence the actual securitisation of the airport, must be understood as the outcome of constant negotiations and dialogue. Only by recognising the particular combination of concerns, goals and benefits perceived by both the police and the airport management can we understand the modalities and use of airport CCTV.

A friendly airport

Besides the role of CCTV in the fight against petty criminality, numerous other camera applications can be found. For the purpose of this chapter, it is particularly revealing to mention the use of CCTV as a means to provide various forms of assistance to the airport clientele. By way of example, consider the following account, relating to the police's contribution in the search for lost luggage and lost children:

> We can also help to find lost luggage. [To do our job], one needs to like responding to people's needs. [...] If we've got time, we like to help, with our telephones, cameras, and even with our control room. In a way, these belong to the people. It's their taxes as well. [...] It's true that we have to deal mostly with thieves, but sometimes, there are people in need of some help. We will also use the cameras in cases of lost children. Whenever there are many people at the airport, we will have many cases of lost children (Airport CCTV operator I, translated by the authors).

In the context of our chapter, this account is interesting because it places CCTV – and hence airport policing more generally – outside a strict risk and security problematic. The quote above implies a definition of airport policing not only as a response to issues of security and risk, but also as a contribution to the effective functioning of the airport. It is not only a crime-free, but also a friendly, airport that is at stake, and in this sense, the police also assumes the role of a watchman, or vigilante, in the service of the airport. As one of our interviewees put it,

> Are they [the police] doing the work of a watchman for a private institution [the airport], or are they limited to public policing operations? I'd say it's a bit of both... (Senior member of technical services at Geneva Airport, translated by the authors).

A 'clean' airport

To further explore the police's contribution to the effective functioning of the airport, another aspect of airport policing, relating to the control and preservation of the airport's house rules, must be emphasised. As the following quotes underscore, airport policing not only aims to reduce criminal behaviour and to assist the airport clientele, but also to control, to prevent and to exclude unwanted behaviour and individuals. Thus the police also contribute to the creation of a 'clean' airport.

> Some time ago, not so much nowadays, young people with roller-skates used to consider the airport as their training ground. In cases we will intervene. Often, it is via telephone [to private security agents], but sometimes we send our own patrol (Head of the airport police control room, translated by the authors).
>
> Sometimes, homeless people find themselves within the airport area. Again, we will ask them to leave the airport. Sometimes we will find them a place with specific associations where they might find accommodation (Airport CCTV operator I, translated by the authors).

The two quotes tell us something about the qualities and meanings of the airport, and about how these are reflected in everyday airport policing. The airport is not to be understood as an open, shared space of public use – on the contrary, access to, and use of, Geneva International Airport is governed by specific house rules and regulations. In accordance with the airport's combined vocation as a national entrance gate and as a commercialised shopping zone, there are some people – and some things – that are not welcome.

Airport policing must be situated within this context. Whilst our interviewees described neither skateboarding youngsters nor homeless people as a security threat, they nevertheless perceived them as disturbing elements to the airport's vocation and reputation. Airport policing here is not only about risk management, but also about the enforcement of house rules and restrictions and, consequently, the exclusion of anything unwanted. As with the previous example, airport policing is here positioned outside a strict risk and security problematic.

To summarise, we can say that airport policing not only aims to create a safe and crime-free airport, but also a friendly and clean airport. From this broad, regulatory perspective, the distinction between the illegal and the unwanted is blurred, since both must be controlled and prevented. This is also the case in the initial dichotomy introduced in this chapter between the police's security concerns and the airport management's business interests. In reality, the relationship between the police and the airport management does not articulate two distinct positions, but two overlapping positions, whose common ground lies in the creation of an attractive and safe 'airport experience' (Aéroport International de Genève, 2008: 30). For both parties, this endeavour takes a turn that goes beyond a strict risk and security problematic.

Only if we recognise these broad, blurring and overlapping objectives aiming at a safe *and* consumer-friendly airport, can we understand the precise modalities and effects of airport policing. Hence the importance of a micro approach that centres on the question of how exactly everyday security practices – and their underlying relationships – are responding to, and shaped by, specific interests and interacting concerns.

Tensions in airport security

Despite the intrinsic combination of interests in the creation of a safe, friendly and clean airport environment, the positions and efforts of the airport management and police can also give rise to tensions and dissonances. In the following, our task will be to provide a series of micro illustrations – some of which have been previously explored from other thematic angles (Klauser, Ruegg and November, 2008) – relating to the existing tensions between the commercialisation and the securitisation of the airport. More specifically, we will offer insights into the difficulties for the airport police in its quest for a safe, ordered and presentable national entrance gate; difficulties which arise from the organisation of special events and performances at the airport.

As mentioned previously, to enhance the 'airport experience' and to attract additional customers, various kinds of special events are organised by the airport management, ranging from displays of model cars to fashion shows, arts exhibitions and flea markets. Whilst some of these events affect the arrival and departure halls, the largest and most 'chaotic' events (such as flea markets and Christmas markets) are held in the airport's railway station.

Airport materialities

The police agents interviewed did not openly disagree with efforts to increase the commercial appeal of the airport. However, special events and performances entail a range of difficulties and complications for the policing of the airport. Regarding airport CCTV operations, for example, a first series of difficulties linked to commercial events is caused by associated changes to the airport's materiality, such as event-related arrangements of objects, placards and decorations.

> Sometimes, we will ask to change the location or positioning of placards, because they limit the cameras' view [. . .]. Recently, the airport workers even put another placard in the middle of the hall, which was not really the best solution for us. We

> also have to pay attention to Christmas decorations. It's very simple; with all these Christmas decorations, we lose a big part of our vision [...]. Recently, we also had to re-position and re-align three cameras because some constructions built by the airport were in our way. Hence we did adapt ourselves to this (Head of the airport police control room, translated by the authors).

The quote underscores the case by case arrangements, adjustments and re-positioning of placards, decorations and CCTV cameras, illustrating the continuous compromises being made between security and business. In this sense, airports are not to be understood simply as spaces of complete control and security, as the current security rhetoric suggests. On the contrary, airport policing results from constantly redefined and rearranged compromises between numerous private and public actors and interests. In these compromises, security does not always trump economy.

Airport uses

Put simply, the safest airport is an empty airport. In reality, however, this is not the police's ambition. On the contrary, as we have seen, the police contribute in many ways to the creation of an attractive and friendly airport. The simplicity of this statement, however, hides a more profound issue, relating to the tensions existing between the commercialisation and the securitisation of Geneva International Airport. For the police, increased numbers of visitors and customers at the airport, attracted by special events and performances, indeed constitute an important challenge. There are at least three reasons for this, each of which deserves some discussion here.

Firstly, unlike normal passengers, event visitors are not passing swiftly through the airport, but remain inside the building for longer periods of time. They cannot be subjected to prescribed patterns of movement, or filtered and controlled through check-in, ticketing, security checks, etc. On the contrary, event visitors follow their own spatial logics, implying complex and opposing micro movements to the general flow of travellers, which heightens the challenge they pose for policing and surveillance operations.

Secondly, according to our interviewees, large numbers of people and distractions are not only more difficult to monitor, but also present ideal conditions for pickpockets and luggage thieves hidden in the crowd. From this perspective, commercially motivated efforts to increase the airport clientele do have a negative impact on the risk of petty criminality at the airport.

> Every special event – I mean whenever there are many people here, bringing money to the airport – is a moment of risk. At such times, pick-pockets will be present (Airport CCTV operator I, translated by the authors).

Thirdly, special events can be difficult to reconcile with increased security precautions against terrorist threats by abandoned luggage. On the one hand, unattended suitcases, boxes or parcels are routinely detected by the police and exploded by specialised security forces. On the other hand, many of the organised events (especially flea markets) are likely to lead to more objects of various dimensions lying around in the airport premises, which will complicate the police's task considerably.

These difficulties reiterate the existing tensions arising from the combined efforts to both secure and commercialise the airport. On a micro level, we find here a powerful illustration of Mark Salter's analysis of the airport as a 'space of shared authority where sovereign and disciplinary powers are both mediated and disaggregated. Within the specific workings of preclearance and airport security, the airport is neither a smooth transit zone nor simply a gate into the nation, but a complex of private and public agencies wrestling with the impossible task of perfect security and perfect mobility' (Salter, 2007: 62). For our investigation of how risk management permeates and shapes particular places and moments, these insights are of high relevance, as they exemplify the variety of factors and interests at work in everyday security and surveillance practices. At this level it becomes apparent just how necessary it is to critically investigate the interdependences, alliances and tensions in contemporary security governance if we are to understand the implications of security and risk for our everyday lives.

The role of external technology suppliers

So far, our study has explored the interactions between two key stakeholders in airport security: the airport police and the airport management. We have chosen this focus because it provides a symptomatic example of the interwoven security and business concerns in contemporary risk management. In reality, however, many other actors, needs and concerns are shaping the securitisation of Geneva International Airport, from the Swiss regulatory bodies in the aviation sector to the shopkeepers in the airport, and from the airport's legal and technical departments to local politicians (Ruegg *et al.*, 2006).

In this chapter, we do not seek to give an exhaustive interpretation of the roles played by all these stakeholders. We do, however, propose at this point to move beyond the issues explored so far, to scrutinise in more detail the position of another key player in the making of airport security: the private supply company of the airport CCTV system. With this extended focus, we hope to bring to the fore another critical issue in contemporary security governance: the increasingly important role played by companies specialising in security and surveillance technologies.

As shown elsewhere in more detail (Klauser, 2009), the use of high-tech surveillance systems considerably extends the role of technical expertise, and the authority of private business companies, in contemporary security governance. If we are to understand how different public and private actors connect in particular locales, special attention must be paid to the providers and designers of the surveillance systems in place.

Technical expertise

A closer look at the history of CCTV at Geneva International Airport reveals that the airport management and police have always had a marked preference not just to purchase CCTV material, but also to buy specialised services and strong client relationships, provided by a highly qualified designer of surveillance systems. As the following quote shows, from the first installation of CCTV equipment at Geneva International Airport, technicalities such as the positioning and location of the cameras have not only been discussed between police and airport representatives, but have also been influenced by the *technical expertise* of private CCTV suppliers.

> When we first meet with suppliers, we may speak about some specific issues of our project. This may then result in a second phase of negotiations. In this case, the supplier might tell us 'I can do what you want but this will cost you four times as much as another solution, which still satisfies 80 percent of your needs'. This level of negotiation not only typically runs in collaboration with the supplier, whose expertise is greater than ours, but also includes the final user of the material. To rely on the supplier's experience is very important for us (Member of technical services at the airport, translated by the authors).

To this account, it is interesting to add the supply company's perspective. As a part of our study, we interviewed the co-founder of *Alarme et Sécurité*, the supplier of additional CCTV cameras at Geneva Airport in 2003.

> Good clients have confidence in us; they are correct in business matters. Bad clients only want to find the cheapest solution. In security matters, the cheapest solution is always too expensive if it doesn't work well. [...] Fortunately, I can do without these bad clients. But we had to manage our company in this sense (CCTV supplier at Geneva Airport, translated by the authors).

The quote offers an additional viewpoint to our discussion so far, in that it reveals the supplier's active quest for responsibility and autonomy in designing 'his own' CCTV system. The external designer and seller of the system is here portrayed as an important stakeholder, harbouring his own vision of the 'right solution', his own form of expertise and his own interests, which are shaping the making of airport security.

Marketing as a source of information

The interests of private businesses in participating in current security matters can be further explored by looking into some of the marketing and selling strategies employed.

> Suppliers generally know that we are the users of the system, which is exactly why they get in touch with us. However, some companies organize larger presentations which are destined for other services as well. In such cases, representatives from other services, including the border police department and the airport, will also join us. [. . .] Recently, somebody came here from a security specialist in Paris, representing several manufacturers of CCTV equipment, in order to present two new camera systems. He was selling a whole series of establishments from door to door. In the morning, he was at the airport; in the afternoon, he was meeting someone somewhere else (Head of the airport police control room, translated by the authors).

This quote is remarkable: not only does it reveal the intense marketing and promotional activities of private companies wishing to take part in the securitisation of Geneva International Airport, but also, more specifically, it portrays these companies as ambulant sellers, presenting and selling more or less standardised solutions for supposedly similar security threats in different places. Indeed, the police and airport representatives interviewed within this study not only associated technology supply companies with specifically commissioned projects, but also acknowledged them as a source of information regarding new solutions available on the market. Thus although such companies are in most cases not practically involved in the setting up and development of particular surveillance solutions at Geneva Airport, their marketing efforts implicitly shape potential future demand by promoting and publicising the latest, internationally established 'security exemplars'.

> I am sure that the technical means which are available to us also create our needs. It's quite trivial really: if we don't have the means, we won't have the action either. From there, to know how tools and techniques evolve is mainly a question of marketing. As users [of surveillance technologies], we won't personally develop anything new to meet our needs. We are much more likely to use and to apply something that's already offered by the market (Member of technical services at the airport, translated by the authors).

In light of the above, the impact of technology supply companies on airport security – and on contemporary security governance more generally – can be seen on at least two levels. Firstly, this example stresses the technical expertise required to manage and to install the airport's CCTV system, which is likely to give the authority of private specialists more weight. Secondly, marketing and promotional activities play an important role in creating demand, and hence in channelling future modalities of security governance. In sum, we see again that security governance is not about the application of universal principles, but a matter of everyday interactions between different actors, with different forms of authority and expertise, each with specific concerns and interests. In the case of ambulant technology suppliers, it is interesting to note that these interests are not anchored in the airport area itself. Rather, technology companies are following their own, de-territorialised business interests. As Holden and Iveson (2003: 66) put it, they are 'wandering the planet in search of consultancy fees and places to save, 'parachuting in' to localities with plans and designs and then moving on to the next place.' Nevertheless, such companies make an important and central contribution to airport security, which deserves therefore critical attention and investigation.

Conclusion: arising challenges for critical risk research

This chapter has provided a set of micro illustrations with regard to the multiple (congruent and conflicting) interests in the securitisation of Geneva International Airport. Airport policing has been positioned within a complex field of needs, driving forces and motivations, bringing together a range of internationally operating stakeholders as well as diverse local actors, predispositions and impulses. More specifically, given our focus on three key stakeholders in airport security – the airport police, the airport management and the company supplying the airport's CCTV system – the chapter has highlighted some of the interdependences, alliances and tensions arising from the simultaneous efforts to create a safe *and* a consumer-friendly airport environment. The examples given have underscored the overlapping security and business concerns in the fight against petty criminality, in the search for lost luggage and children, and in the control of disruptive behaviour. They have also, on the other hand, underlined the delicate balance between security and business needs, with respect to the material arrangements of the airport (for example in the positioning of placards, decorations and CCTV cameras) and in view of the problems arising from the staging of sales-stimulating events.

These insights are of exemplary value in answering the question of how exactly security governance permeates particular places and moments. We need to recognise not only the blurring and overlapping concerns in contemporary security governance, but also the multiple tensions and dilemmas arising from the ways in which specific problems are framed, approached and exploited for particular needs. It becomes clear, therefore, that risk and security issues are not pre-given or value-free, but shaped by complex relationships and interactions bringing together various actors and interests.

Challenges for critical risk research

To conclude, it is worth pointing towards three basic challenges, or avenues for further critical investigation, that emerge here for future research on issues of risk and security.

Firstly, there is a pressing need for more detailed empirical research into the complex relationships between the various institutions and agents which shape the everyday micro-politics of risk and security. In this chapter, we do not claim to provide an exhaustive interpretation of all the factors at stake. This study has outlined some of the interacting concerns (and some of the resulting effects) in security governance, but much more detailed and comparative empirical investigations are needed in order to scrutinise the various interests, logics and impacts of contemporary security operations. Furthermore, it will be of major importance to seek more elaborate insight into the micro-scale implications of security practices on the everyday life of individuals and social groups. How are communities affected by security procedures, operations and strategies? And what types of interests and relationships lie behind these interventions?

Secondly, to be truly credible in this research ambition, critical risk research must also critically interrogate its own underlying values and objectives. Just as security governance is inherently complex and dependent on a wide range of interests, a study of risk is also mediated by specific norms, intentions, institutions and agents, which directly and indirectly shape the form, direction and content of the outputs produced. In this sense, the insights provided in this chapter with respect to the interacting concerns, forms of expertise and coalitions of authority in security governance also serve as a reminder of the variety of factors at work in the production of knowledge through academic research in risk and security matters. What is needed, therefore, is an attitude of suspicion, a kind of self-sceptical reflex, which leads to a constant questioning of the values, driving forces and power relationships underlying the study of risk and security. What aims, ambitions and socio-political and institutional environments characterise the study of risk and security? In what ways, and to what ends, are problems of risk and security framed and studied? For whom are insights provided and solutions produced?

Thirdly, and following directly from the above, critical risk research is not only situated within a complex grid of relationships of power. Instead, the field of critical risk research itself – by its practices and knowledge, and by the interventions it generates – also participates in the co-production of 'risk' as a series of problems to manage and to solve. Thus critical attention must also be paid to the impact of risk research on its very object of study. The different forms of expertise studied in this chapter – the airport police's *practical expertise* in the fight against petty criminality, the CCTV supply company's *technical expertise* in the design of surveillance systems and the airport management's expertise in business matters – were each in turn shown to contribute to airport security. Likewise, the information and expertise produced by academic research also influences the sociopolitical processes and practices of managing risk. Paying close attention to the wider social, political and ethical implications of the study of risk and security is hence of critical importance.

In sum, the study of risk and security requires an appreciation of the multitude of contributory causes, interests and actors assembled around the problems at hand. It also raises a series of important questions with respect to the underlying values and hidden *a priori* in this field of study, and to the interventions produced by the insights provided. It is on this basis that the power of critical risk research, and its resulting responsibilities, become evident and questionable.

References

Adey, P. (2004) Surveillance at the Airport: Surveilling Mobility/Mobilizing Surveillance. *Environment and Planning A*, **36**, 1365–1380.

Aéroport International de Genève (2008) *Rapport Annuel 2008*, Geneva.

Aéroport International de Genève (2009) *Rapport Annuel 2009*, Geneva.

Beck, U. (1998) Politics of risk society, in *The Politics of Risk Society* (eds J. Franklin), Polity Press, Cambridge, pp. 9–22.

Castells, M. (1996) *The Rise of the Network Society – Volume 1: The Information Age: Economy, Society and Culture*, Blackwell, Oxford.

Coleman, R. (2004) Reclaiming the streets: closed circuit television, neoliberalism and the mystification of social divisions in Liverpool, UK. *Surveillance and Society*, **2**(2–3), 145–160.

Etat de Genève (2001) *Règlement relatif à la police de sécurité internationale*, 13 June 2001, Geneva.

Franzen, M. (2001) Urban order and the preventive restructuring of space: the operation of border controls in micro space. *Sociological Review*, **49**(2), 202–218.

Fuller, G. (2003) Life in transit: between airport and camp. *Borderlands E-journal*, **2**(1), http://www.borderlandsejournal.adelaide.edu.au/vol2no1_2003/fuller_transit.html. Accessed 28/05/08.

Holden, A. and Iveson, K. (2003) Designs on the urban: New Labour's urban renaissance and the spaces of citizenship. *City*, **7**(1), 57–72.

Klauser, F., Ruegg, J. and November, V. (2008) Airport surveillance between public and private interests: CCTV at Geneva International Airport, in *Politics of the Airport* (ed. M. Salter), University of Minnesota Press, Minneapolis, pp. 105–126.

Klauser, F. (2009) Interacting forms of expertise in security governance: the example of CCTV surveillance at Geneva International Airport. *British Journal of Sociology*, **60**(2), 279–297.

Klauser, F. (2010) Splintering spheres of security: Peter Sloterdijk and the contemporary fortress city. *Environment and Planning D: Society and Space*, **28**(2), 326–340.

Koskela, H. (2000) The gaze without eyes: video-surveillance and the changing nature of urban space. *Progress in Human Geography*, **24**(2), 243–265.

Lyon, D. (2003) Airports as data filters: converging surveillance systems after September 11th. *Information, Communication and Ethics in Society*, **1**(1), 13–20.

Lyon, D. (2004) Surveillance technologies: trends and social implications, in *The Security Economy* (ed. B. Stevens), OECD, Paris, pp. 127–148.

Ruegg, J., Flueckiger, A., November, V. and Klauser, F. (2006) *Vidéosurveillance et risques dans l'espace à usage public: représentations des risques, régulation sociale et liberté de mouvement*, CETEL publication No 55, Travaux du CETEL, Geneva.

Salter, M. (2007) Governmentalities of an airport: heterotopia and confession. *International Political Sociology*, **1**, 49–66.

Salter, M. (ed.) (2008) *Politics of the Airport*, University of Minnesota Press, Minneapolis.

Sorkin, M. (ed.) (1992) *Variations on a Theme Park: The New American City and the End of Public Space*, Hill and Wang, New York.

Stevens, B. (ed.) (2004) *The Security Economy*, OECD, Paris.

Töpfer, E. (2005) Jeden Bahnhof erfassen. *heise online*, 31 August, http://www.heise.de/tp/r4/artikel/20/20832/1.html. Accessed 12/12/2011.

CHAPTER 6
Governing Risky Technologies

Phil Macnaghten[1] *& Jason Chilvers*[2]

[1]Department of Geography, Durham University, Durham, UK

[2]School of Environmental Sciences, University of East Anglia, Norwich, UK

In this chapter we examine debates surrounding the governance of risk, focusing in particular on the issues, challenges and responses related to the governance of new and emerging technology. We start by discussing why governing risky technology is of such relevance today. Having set out the issue, we then discuss different institutional responses, examining how science policy has shifted its rhetoric, from traditional approaches based on a paradigm of sound science, expert advice and one-way communication to an emergent paradigm, characterised by an emphasis on dialogue, openness, transparency and responsible innovation. The UK Sciencewise Expert Resource Centre (Sciencewise-ERC) is introduced as a case study of the new scientific governance in practice. Informed by analysis of recent public dialogue initiatives, we discuss the social and ethical issues that new science and technology introduces into the social domain and that a governance agenda has to grapple with. Finally, we discuss experiments in governance that are being trailed in the UK and Europe in particular technological domain areas, evaluating their efficacy and potential as responses.

Our chapter is set in the context of developments in critical risk research over the latter half of the twentieth century, from a field centred on quantitative risk assessment to one that has increasingly acknowledged the socio-cultural dimensions of risk analysis, public understanding, and institutional response. An important implication of this has been moves to more participatory, open and inclusive conceptions of risk governance (Renn, 2008) and the use of deliberative or survey-based 'technologies of elicitation' (Lezaun and Soneryd, 2007) to understand better public reactions and 'social concerns' relating to emerging technologies and associated risks. These emerging forms of deliberative risk governance have predominantly been imagined by risk researchers, and implemented in practice, as discrete one-off exercises that directly inform policy-making. In response

Critical Risk Research: Practices, Politics and Ethics, First Edition.
Edited by Matthew Kearnes, Francisco Klauser and Stuart Lane.

to issues of public trust, an implicit assumption of this approach has been that public controversies over emerging technologies can be anticipated and the politics surrounding them can somehow be contained. Deliberative forms of risk governance have also been questioned for narrowly focusing on the consequences, in terms of risk and safety, rather than the social and ethical implications of, and human needs and purposes that drive, emerging science and technology (Wynne, 2005).

Throughout this chapter we take these critical insights of the emergent risk governance discourse and attempt to open it up in two main ways. First, rather than viewing public dialogue experiments as one-off events whose primary function is to find out what the public think about a particular technology as a discrete decision input, we stress the importance of mapping across individual dialogues to examine underlying governance concerns that drive public responses. In doing this we show how so-called 'upstream' or anticipatory questions about the purposes, direction, control, and governance of emerging technologies (Wilsdon and Willis, 2004) are also important in honest broker or issue advocate engagement situations more readily associated with 'downstream' or 'risk governance' domains (Felt and Wynne, 2007). Second, we argue that responding to questions of trust and governance concerns held by the public is not simply a matter of communicating better or involving more. It requires a broader appreciation of the science governance system in which public dialogue forms only a part, including the diversity of routes through which public values can shape science and technology as well as modes of public accountability, scrutiny and transparency. Our observation that governance responses are often out of step with public concerns highlights the need for risk researchers to understand better the processes that mediate institutional response, and also actively engage in catalysing more reflective, relational and transformative forms of institutional learning.

Models of governance

Until recently it was assumed that new science and technology were relatively unproblematic public goods, that should be permitted unless and until there were clear and demonstrable indications of harm, either to the environment or human health, identified and preferably quantified through science. This 'risk-based' approach to scientific governance became deeply embedded within science policy practice and culture during the latter part of the twentieth century across western democracies, reproduced through mechanisms of science-based risk assessment, tightly-defined expert advisory structures and a narrow reliance on sound science and administrative caution. In addition, it became coupled to an optimistic and largely modernist social imaginary, where the progress of science was in large measure tied to the future prosperity of the nation.

However, this model of risk governance failed to accommodate a series of political contestations around particular technologies, notably the development of civil nuclear power, where risk-based estimates of harm failed to encapsulate the social, ethical and political stakes associated with technoscientific progress. The model came to be challenged, initially by the academic community and later by wider policy institutions, and for three reasons. First, it became clear that the narrative of technology as a liberating and empowering force was not unproblematic. Just as dreams can turn to nightmares, the promise and hope of technology can equally turn to failure, breakdown, crash and disappointment (Jasanoff, 2006). Indeed, as historians and sociologists of technology have testified, the twentieth century is besieged with examples of technology's dark side, from the horrors of chemical weapons in the First World War to nuclear bombs in the Second, from nuclear explosions at Chernobyl to chemical explosions at Bhopal, from the terrorist atrocities of 11 September 2001 to the on-going perils of climate change, ozone depletion and other environmental disasters (Jasanoff, 2003; Perrow, 1984). While these cases clearly require different kinds of social, political and historical explanation, they nevertheless, in Sheila Jasanoff's words: 'have served collective notice that human pretensions of control over technological systems need serious reexamination' (Jasanoff, 2003: 223). The proposition that *systems* of technology may have the capacity to disrupt society through the manufacture of risk also lies at the heart of a substantial body of sociological literature on reflexive modernisation and the risk society (Beck, 1992, 2000, 2009; Beck *et al.*, 1994). Within this sociological canon, risks are seen as becoming global, unpredictable, distributed in time and space, invisible to the senses and endemic to the ways in which modern life is conducted in technologically-intensive society. Under these conditions, a quest for risk reduction through narrow reliance on sound science and case-by-case scientific risk assessment is by definition a partial and restricted response to life in the risk society.

Secondly, it became apparent as an empirical fact that many of the most profound technologically-induced risks that are faced today were not identified in advance by formal processes of risk assessment. Rather, they appeared as unforeseen surprises, characterised by interactions between unknown processes and variables. In the case of chlorofluorocarbons (CFCs) for example, risk assessment was limited to questions of human toxicity, and where the vertical transport of CFCs into the upper atmosphere was not imagined as a relevant variable (Hoffmann-Reim and Wynne, 2002). Nobody had suspected a connection between stratospheric CFC concentration and stratospheric ozone concentration. According to Wynne, the problem lies with the implicit assumption that risk analysis can identify the relevant interactions and processes in advance (Wynne, 2002). More accurately, risk analysis needs to be recognised as fundamentally limited by ignorance, by its ability to focus on known uncertainties rather than on unknown uncertainties. Thus, the fundamental risk governance question for Wynne concerns who will be responsible for future unanticipated surprises, given that 'unanticipated effects of novel technologies are not just possible but probable' (Hoffmann-Reim and Wynne, 2002: 123).

And thirdly, it became increasingly recognised that assessment of risks are rarely based on science or technical evaluation alone, but are contingent on social judgement and values. To determine the level at which a technology can be deemed safe, or whether a certain level of uncertainty is acceptable, can no longer be read as purely a technical matter; it depends on such considerations as norms, expectations, the trustworthiness of responsible institutions, the purposes to which the technology will be addressed, and so on. The importance of 'framing' as a necessary element in the risk assessment process was advocated in forms of technology appraisal (Grove-White *et al.*, 2000; Stirling, 1998; Wynne, 1992, 2006), and recognised institutionally by the Royal Commission on Environmental Pollution in their 'Setting Standards' report (RCEP, 1998), as a key requirement for assuring public trust in environmental regulation. One implication that follows concerns the need to consider and take into account, at the stage of the definition of the issue, the 'perspectives and values of all those who may be affected by a problem or have an interest in it' (RCEP, 1998: 119). This appeal to include a broader array of voices, including lay publics, in the deliberative process is a major departure in risk governance and will be considered further below. A further implication concerns the need to understand the appropriate role of science in the policy process. Following analysis of the ways in which science has been used in the service of particular interests, Roger Pielke in his book *The Honest Broker* differentiates four distinctive ideal types: the pure scientist, the science arbiter, the issue advocate and the honest broker of policy alternatives (Pielke, 2007). In brief, Pielke argues that in risk domains where there exists high levels of scientific uncertainty and/or low levels of values consensus (such as in relation to controversial questions of climate change, radioactive waste disposal, forest management, air and water pollution etc.), science can offer little to resolve value differences and resolve political dispute. Indeed, when cast in this role scientists may unwittingly aid in transforming a scientific debate into a political controversy. More productively, Pielke advocates the role of the scientist as the honest broker of policy alternatives, who seeks to expand the scope for choice for decision-makers by contributing to the creation of new choices, new problem framings and new decision pathways.

These largely academic critiques gained policy traction in the face of a series of high profile risk controversies and the failure of traditional models of scientific governance to anticipate and take into account adverse public reaction. In the UK and Europe, the controversy surrounding mad cow disease and the uncertainties surrounding the link between bovine spongiform encephalopathy (BSE) and Creutzfeldt-Jacob disease (CJD), followed by the rows over genetically modified (GM) foods and crops in the late 1990s, led to a number of influential policy reports calling for a new style of scientific governance. This included: more proactive public involvement and deliberation, a need to ensure that decisions are based on public values, and wider openness and accountability in the scientific and regulatory process (RCEP, 1998; House of Lords, 2000; HM Treasury, 2004). Embedded within the new scientific governance are a number of assumptions: that current levels of mistrust are a matter of concern for scientific governance and that openness and accountability can remedy it; that expert and scientific framings of risk issues are often at odds with lay ethical judgements and values and that this can be remedied by giving lay people a greater voice; and that public dialogue should be promoted by government as part of the democratic process (Irwin, 2006). The framing of public dialogue itself evolved throughout the early 2000s, from a set of debates around the framing of regulatory science, to a wider and more inclusive set of debates about the social and ethical dimensions of science and technology, and on how these could be anticipated by techniques of public engagement (see Chilvers, 2008; Wilsdon and Willis, 2004).

The Sciencewise Expert Resource Centre

In the UK, the Sciencewise Expert Resource Centre (Sciencewise-ERC), funded by the Department for Business, Innovation and Skills, has been an integral part of the new scientific governance, commissioning and championing public dialogue since 2005. Its birth followed the demise of the Copus Grants Committee, a funding body administered by the Royal Society and charged with bringing scientific issues to the attention of the wider community and promoting a more scientifically literate society. The idioms of Copus lay in one-way forms of science communication and deficit models of public understanding. In its place, Sciencewise (renamed Sciencewise-ERC after 2007) spoke to the new discourse of two-way dialogue and public engagement. Its mission was avowedly more strategic, placed within the forward-looking Department of Trade and Industry and with overt ambitions of using dialogue to inform policy.

Thorpe and Gregory (2010) identify Sciencewise-ERC as part of a policy convergence between knowledge production and wealth creation. They argue that Sciencewise-ERC activities can be seen as part of the production of Post-Fordist publics, providing the State with intelligence to ensure the smooth passage of incoming-generating science and technology. A related argument has been promoted by Wynne (2006) who has criticised the institutional use of public engagement as a means to restore public trust in science. However, while instrumental logics can be attributed to institutionalised forms of public engagement other logics are manifest too. Disaggregating the latter point, Stirling (2005) has analysed the various motivations that underpin government-funded public dialogue. These include: the normative (e.g. that dialogue is the right thing to do for reasons of democracy, equity, equality and justice), the instrumental (e.g. that dialogue provides social intelligence to deliver pre-committed policy objectives, such as those of building trust or of avoiding adverse public reaction), and the substantive (e.g. that policy choices can be co-produced with publics in ways that authentically embody diverse social knowledge, values and meanings in a substantive manner).

In substantial terms, Sciencewise-ERC has constructed and guided a number of dialogue projects, from nanotechnology and stem cell research to the use of DNA in forensics and building low carbon communities. Using Pielke's terminology, the aim of the dialogue is to open up the competing values, ethics, issues and interests associated with new science and technology, such that this can provide social intelligence to policy-makers and help inform choices and trajectories. The Sciencewise-ERC has in addition fulfilled a normative role in encouraging the appetite for public dialogue within government and its agencies, through the provision of opportunities for training and networking, as well as through guidance on best practice. The Sciencewise-ERC thus can be seen as a central platform in the UK Government's broad ambitions for a society in which the public, the broad science community and policy makers feel comfortable with the direction taken by science and technology. Other parallel initiatives include the development of the *Beacons for Public Engagement* (HEFCE, 2007) together with greater funding and support from the research councils and HEFCE to help embed public engagement in science within universities. There have been in addition new governmental initiatives, such as *Science for All* (Science and Trust Expert Group 2010a) and *Science and Trust* (Science and Trust Expert Group 2010b), which have attempted to encourage new forms of institutional engagement – including a call for a 'public compact' on cross cutting issues of science and technology, resulting in the *Concordat for Engaging the Public With Research* (RCUK, 2010), launched in December 2010.

Such initiatives are part of a growing shift in institutional rhetoric and practice from a focus on public dialogue and engagement (as an end in itself) in responding to issues of public trust in science, towards initiatives where public dialogue is integrated into the governance of science and innovation (see Felt and Wynne, 2007). However, before examining the efficacy of governance mechanisms to respond to the issues presented by new science and technology it is necessary to articulate and describe these issues in detail. Our approach to articulating these issues is through analysis of the science governance issues and concerns expressed by lay publics in the 17 public dialogues co-funded by Sciencewise-ERC. The rationale is threefold: (1) the 17 dialogues represent a unique body of data on public attitudes to new science and technology that is unparalleled in a western context; (2) understanding cross-cutting themes enables reflection on the factors that shape public enthusiasm and concern; and (3) surprisingly such analysis has hitherto escaped the academic literature. This analysis is informed by a review of public concerns, governance and institutional response undertaken by the authors for Sciencewise-ERC (see Chilvers and Macnaghten, 2011).[1]

Introducing the dialogues

The 17 dialogue projects on science and technology that Sciencewise-ERC has co-sponsored since 2005 relate to a broad spectrum of different science, from the biosciences to climate change, from healthcare to information technology, from environment to nanotechnology (see Appendix 1 for a full list). The range of dialogues is intended not simply to cover a broad spectrum of issues associated with novel science and technology but also designed to inform the policy process. Thus, with the exception of one (*Science Horizons*), each of the dialogues has been co-sponsored by an external government department, agency, third sector body or company. However, it is clear that while the dialogues have been designed to inform the policy process, they do so in distinctive ways and with different underlying theories of change. Informed by Pielke's typology as set out earlier, in which he distinguishes between different models of scientific advice in the

[1] The review was undertaken as part of a project on behalf of Sciencewise-ERC titled 'Science, Trust and Public Engagement'. Led by TNS-BMRB, in partnership with the Royal Society and researchers at the Universities of East Anglia and Durham, the project aimed at understanding the institutional context of science governance in the UK, its responsiveness to public concerns and ways of promoting pathways to good governance. The source material for the review included: reports on the Sciencewise-ERC dialogues and associated evaluations, reports and guidance notes from the Sciencewise-ERC resource library, evidence emerging from the evaluation of the Beacons for Public Engagement, evidence from the ESRC Critical Public Engagement Seminar Series, and associated academic literature on public engagement and science and technology studies.

policy process (Pielke, 2007), we have differentiated three distinct models of public engagement.

Upstream Model

The aim of the *Upstream Model* of public dialogue is to develop a process through which publics can engage in complex conversations on the range of issues and questions posed by science and technology at a relatively early stage in the innovation process. The conversations tend to be exploratory, to concern the ways in which the science and technology is being imagined by social actors, to scrutinise the views and visions of actors, to articulate what are the social and ethical issues, and to deliberate on the factors that shape concern, hope and possibility. Such dialogue events tend to be tied only loosely to specific policy goals and outcomes, and are more apt to offer generic advice on the governance of science and technology. In Table 6.1 we identify dialogue projects that use the *Upstream Model*. These include projects on specific areas of emergent science and technology (*Synthetic Biology, Nanodialogues, Industrial Biotechnology, Geoengineering*), as well as projects that explore possible future directions in science in general (*Science Horizons, Community X-Change*). The *Industrial Biotechnology* and *Community X-Change* dialogues also fall within the *Issue Advocate Model* since the framing of both subscribe to a particular policy goal: that of building confidence in industrial biotechnology and that of providing a voice for local communities respectively. The *Geoengineering* dialogue also falls into the *Honest Broker Model* since the context in which the discussion took place related to a very specific policy question: that of whether, and under what conditions, *Geoengineering* approaches have a role to play in tackling climate change.

Honest Broker Model

The aim of the *Honest Broker Model* of public dialogue is to deliberate on different policy options and to determine a preferred policy outcome, with justification. Dialogues that fall under this model tend to position the role of the public as a lay ethical arbiter in providing views on how to proceed, weighing up the pros and costs of different courses of action, articulating the conditions under which different options are acceptable or not, and thus helping inform a policy decision. Such dialogue events thus tend to be tied closely to specific policy goals and outcomes, on domains of science and technology which are relatively developed, and which are known already to pose social and ethical problems and dilemmas. In Table 6.1 we see that the *Honest Broker Model* tends to be favoured for health-related questions, where the aim is one of helping sponsors to determine whether, and under what conditions, to fund and move forward with specific and ethically-challenging research (*Animals containing Human Material, Hybrids and Chimeras, Stem Cells*). A related category of dialogue relates to projects

Table 6.1 Sciencewise Dialogue Activity by Model of Public Engagement

	Model of Public Engagement		
Dialogue Project	Upstream	Honest Broker	Issue Advocate
Synthetic biology	√		
Industrial biotechnology	√		√
Community X-Change	√		√
Big energy shift			√
Low carbon communities challenge			√
Geoengineering	√	√	
Energy 2050 pathways			√
Landscape and ecosystem futures		√	
Science horizons	√		
Risky business			√
Animals containing human material		√	
Stem cells		√	
Hybrids and chimeras		√	
Drugsfutures		√	
Trustguide			√
Forensic use of DNA		√	
Nanodialogues	√		

that seek to determine the effects of advances in the sciences on issues that already pose social problems and dilemmas (*Drugsfutures*) or that pose new dilemmas for the use and management of information (*Forensic Use of DNA*). The final example is tied to policy on future land use planning in the context of climate change (*Landscape and Ecosystem Futures*), and where the role of the dialogue was to understand values, benefits and trade-offs in relation to future land use scenarios.

Issue Advocate Model

The aim of the *Issue Advocate Model* of public dialogue is to deliberate on a shared policy goal, such as galvanising community-led participation in climate change or building trust in a particular science or technology, and to deliberate on the conditions under which this goal will be realised. Dialogues subscribing to this model are characterised by adherence to a pre-determined policy goal, and where the aim of the dialogue is to develop new ways to achieving that goal through better understanding the views, beliefs and needs of publics. In Table 6.1 we see that the *Issue Advocate Model* is common in dialogues on climate change, responding to the policy goal of stimulating behaviour change as part of the Government's commitment to a legally binding target of at least an 80% cut in greenhouse gas emissions by 2050. Policy-derived questions shaping the various dialogues include: how to cut emissions at a local level (*Low Carbon*

Communities Challenge), how to encourage people to change their energy behaviour (*The Big Energy Shift*), how to engage representatives of the public in local communities across the UK to run local deliberative dialogues (*Energy 2050 Pathways*), how to increase awareness in students aged fourteen and over of issues of risk in science (*Risky Business*), and on how to improve cyber trust (*Trustguide*).

Common themes and challenges for governance

There has been little attention drawn to the common themes of concern that have emerged across public dialogue initiatives, including the Sciencewise-ERC dialogues, and of their cumulative significance for questions of governance (for an exception see Start, 2010). In this section we analyse key cross-cutting themes shared across the Sciencewise-ERC dialogues, and discuss their relevance and meaning in relation to wider public engagement literature and practice.

Survey evidence provides a snapshot understanding of public attitudes. Since 2000, the UK Government have commissioned four surveys looking at the UK public's attitudes to science, scientists and science policy. Conducted in 2000, 2005, 2008 and 2011, they represent the Government's main mechanism for assessing progress on public engagement with science. The recent UK Public Attitudes to Science (PAS) 2011 survey reported general enthusiasm for science and scientists (Ipsos MORI, 2011). Seventy-nine per cent of respondents agreed that '*on the whole*, science will make our lives easier', 54% that 'the benefits of science are greater than any harmful effect', 88% that 'scientists make a valuable contribution to society', while 82% agreed that scientists 'want to make life better for the average person'. The Sciencewise-ERC dialogues reinforce this picture. Across the dialogues we are presented with a public who is generally positive, upbeat and excited about science, and about its transformative potential in delivering improvements to our everyday lives and to the environment.

However, while the PAS 2011 survey reported positive attitudes in general there remained concerns about the future developments of science: 40% agreed that 'the speed of development in science and technology means that it cannot be properly controlled by Government'; 56% that 'people shouldn't tamper with nature'; 54% that 'rules will not stop scientists doing what they want behind closed doors'; while 30% agreed that 'scientific advances tend to benefit the rich more than they benefit the poor'. In addition, attitudes to science depended on which technology sector is being discussed, with health-related and public-interest science generally more favoured than previously controversial science such as nuclear and genetically modified crops. This latter point complements qualitative research that suggests that public attitudes towards science cannot easily be segmented into those that are pro- and anti- science, or even more fine-tuned clusters adopted by the PAS 2011 survey (i.e. the concerned, the indifferent, the late adopters, confident engagers, distrustful engagers and disengaged sceptics). It suggests that a wider set of contextual factors are in play and that closer attention needs to be paid to these to understand the ambivalent responses that characterise public responses to science (for analysis of ambivalence in public risk research, see Grove-White *et al.*, 1997; Kearnes and Wynne, 2007; Kearnes *et al.*, 2006a; Macnaghten and Urry, 1998). In and across the Sciencewise-ERC dialogues, five contextual factors were analysed to be central to the structuring of public attitudes, and are discussed below.

Purposes

A key factor shaping people's attitudes towards science and technology concerned their assessment of the purpose of the science, and of the motivations of those involved. In whose interests is the science being developed? Are particular innovations necessary? Are there alternatives?

Medical and health technologies were seen by and large as driven by good purposes, including that of curing diseases, improving wellbeing and prolonging life. Research was thus accorded high importance, even when there were acknowledged ethical dilemmas. For these reasons, the public supported their role as *Honest Broker* in health-related dialogues (*Animals Containing Human Material, Hybrids and Chimeras, Stem Cells*), helping to determine whether, and under what conditions, to fund and move forward with specific and ethically-challenging research. In the *Stem Cells* dialogue, support for further advances in the science was seen as conditional on the purposes of the research and on whether it respects human values: Would it reflect public rather than solely commercial interests? Would it respect individual rights and autonomy? Was it focusing on serious diseases? Were people involved in decision-making processes? While in the *Animals Containing Human Material* dialogue, support was similarly premised on the assumption that the aims of the research would be to improve human health or cure human diseases. For *Upstream* dialogues (*Science Horizons, Synthetic Biology, Nanodialogues, Industrial Biotechnology*), the question of the purposes to which science would be directed was also a central issue. In the *Synthetic Biology* dialogue, for example, the motivations of scientists were deemed to be a key determinant for assuring acceptability: What is the purpose of your research? Why are you doing it? What are you going to gain? What else will it do? How do you know you are right? Given that the science was at an early stage, with clear potential for good and bad, ensuring that the science was conducted for good reasons (i.e. in response to societal needs rather than for short-term gain or for knowledge for its own sake) was presented as a critical question.

The picture that emerges from the dialogues is that while there is a general belief that science and scientists are motivated by the common good, supporting the findings of the PAS 2011 study, this level of trust depends on the science being seen as conducted for good reasons. These are whether the science is directed to societal rather than to commercial goals, with the goals of curing disease and improving health clearly the most favoured; and whether the science itself respects basic human values.

Trust

A pervasive feature across a number of dialogue projects was that people rarely trusted the motives of Government to act in the public interest. This general mistrust was prominent in *Upstream* dialogues (*Synthetic Biology, Nanodialogues, Industrial Biotechnology*), where even at an early stage of innovation R&D, the direction being undertaken by science was seen as in danger of being overly directed by private rather than public interests. This distrust is apparent especially in domains where there is a perceived proximity between government and industry, most notably in agricultural and industrial biotechnology. As the report on the *Industrial Biotechnology* dialogue states: 'Profit or anything associated with industry are viewed with great suspicion and there is little faith that the Government will effectively resource the control and monitor[ing of] industry' (p. 4). Indeed, while the motives of scientists may be trusted, in general, the motives of government and industry are not. As Start comments in his review: 'The initial public perception of government is of a regulatory structure that is weak and unreliable, vulnerable to private interests, and vulnerable to dangerous products slipping through the net' (Start, 2010: 20).

Trust was also an issue in *Issue Advocate* dialogues, notably on climate change issues where the science was presented as being conducted for good purposes (e.g. to save the planet) and where these claims were disputed in the face of real world contingencies. In the *Big Energy Shift* dialogue, for example, participants expressed a lack of trust that either the Government, or the private sector, would actually deliver a fair and equitable system to respond to climate change. In the *Honest Broker Geoengineering* dialogue, there was similar mistrust expressed in the accuracy and robustness of climate models and projections, and in the extent to which scientific research can ever truly be independent of the interests and agendas of funding agencies. The notable exception was on health-related *Honest Broker* dialogues where, alternatively, there appeared to be an underlying sense of trust and confidence in regulation, oversight and in the good intentions of Government (*Animals containing Human Material, Hybrids and Chimeras, Stem Cells*).

Unfortunately there has been little analysis across the dialogue reports of the reasons that underpin such generic public distrust in government, or advice on what would constitute an appropriate governance response, aside from appeals to improved communication of benefits (*Industrial Biotechnology*), for regulation to develop an anticipatory and social intelligence capacity (*Synthetic Biology*), for more attention to be given to the equity effects of scientific and technological options (*Geoengineering*), and for the need for more inclusive and systematic dialogue (*various*). A particular absence is an account of why science and technology has become a site for political mobilisation in some spheres (e.g. biotechnology and food) and not others (health), of the reasons why Government is seen as not to be trusted to monitor industry or look after the long-term, and of the need to differentiate between systemic as opposed to localised forms of mistrust.

Powerlessness

In 1995, in a project on public perceptions and sustainability, Macnaghten *et al.* (1995) observed that the pronounced fatalism and cynicism that people expressed towards national and local government was a key barrier to environmental behaviour change. They further argued that attempts by government to galvanise community action would depend on their ability to develop relational mechanisms through which a sense of inclusion and shared purpose could be established. Ten to fifteen years later, and despite a heightened institutional rhetoric on inclusion, it is clear that many people still feel they are not included in deciding what kinds of public science and technology gets funded and for whose interests: i.e. they feel 'kept in the dark'. This sense of powerlessness is expressed well in an extract from the *Synthetic Biology* dialogue report:

> There was a strong sense that scientists are a closed community – while research was scrutinised by peers, it was hard to access by others. In part, this was because scientific expertise and knowledge of a field set them apart from others. However, it was also believed there was a cultural resistance to opening up science to the views and values of the public. This was particularly problematic as participants felt compelled to trust scientists, but ultimately felt powerless to have any control. As one participant noted: *'How can I stop a whole team of scientists doing something? I feel I can't, I feel powerless. (Female, AB, 18-34)*
>
> (*Synthetic Biology*, 2010: 41*)*

Similar views were expressed in other *Upstream* and *Issue Advocate* dialogues: in the *Industrial Biotechnology* dialogue participants expressed a sense of disaffection from the scientific and industrial process; in the *Big Energy Shift* dialogue, individuals were seen as having little agency in promoting behaviour change; and in the *Nanodialogues* people felt similarly disaffected, as indicated in some of the dialogues, where one participant commented, tongue in cheek, on the peculiarity of the dialogue process: 'I feel lucky, I feel like we can make some nanoscule contribution to society'.

Participants responded to this acknowledged sense of powerlessness with a number of overlapping suggestions for increased citizen participation and engagement. At the downstream end, where products and technologies are already in the market place, came suggestions for the provision of improved, balanced and honest information: on harms (*Drugsfutures*), on rights (*Forensic Uses of DNA*), and on the provision of guarantees in the event of mishaps (*Trustguide*). With both upstream and downstream science, came calls for investment in risk and safety research: for research on effective governance and quality control procedures (*Stem Cells*), on long-term monitoring for unpredictable effects (*Nanodialogues*), and on the containment, control and governance of biotechnology risks (*Industrial Biotechnology*; *Synthetic Biology*). At the more upstream end, where the risks of the technology were uncertain and less understood, were calls for more open discussion of the uncertainties in the science and their potential effects: within the medical establishment (*Drugsfutures*), the judiciary (*Forensic Uses of DNA*), and companies (*Industrial Biotechnology*). A common theme was the call for a change in the culture of science that would encourage scientists to voice concerns over potential risks and uncertainties, and to reflect on wider social and ethical considerations. Thus, while participants expressed a sense of powerlessness across all models of dialogue activity, across the *Upstream* dialogues powerlessness tended to be expressed as fatalism (an attitude of resignation in the face of a political economy of science that they have minimal power to shape), in the *Issue Advocate* dialogues powerlessness was more likely to be expressed as scepticism (an attitude of doubt to the apparent claims of official institutions), while in the *Honest Broker* dialogues, powerlessness tended to be an issue only so far as public interest criteria were seen as unlikely to inform policy decisions.

Speed and direction

A long-standing public concern is that research and innovation processes are being developed at a speed that exceeds their scope for ethical and regulatory oversight (for an academic treatment of speed, see Bingham, 2008; Stengers, 2000). In the PAS 2011 survey for example, 40% of respondents agreed that 'the speed of development in science and technology means that it cannot be properly controlled by Government', an increase of 4% over the 2008 survey. Examples of this kind of concern can be seen in the *Stem Cells* dialogue (Was research being pushed to deliver applications too soon?), and in the *Synthetic Biology* dialogue (What were the dangers of speeding up natural and evolutionary processes?). As can be seen from the examples above there were two variants. In the *Honest Broker* dialogues this concern tended to be expressed through the danger of short-term commercial pressures trumping social and ethical considerations, while in the *Upstream* dialogues this concern was expressed more ontologically, in relation to the power of emergent science to disrupt and mess with natural orders and processes (for an extension of this argument, see Macnaghten, 2010).

Concerns were also voiced on the direction science is taking us, and whether this has been adequately considered and deliberated upon in advance (see also Stirling, 2007). These concerns, chiefly manifest in the *Upstream* dialogues, extended beyond matters of safety and technical risk to a broader set of social and ethical issues that included: concerns over unforeseen consequences including controllability and reversibility (*Geoengineering, Synthetic Biology*), impacts on perceived naturalness (*Geoengineering, Synthetic Biology*); and impacts in terms of fairness and equity (*Geoengineering*). As the *Nanodialogues* report commented: 'Safety was a sideshow. The real concern was with where companies are taking us' (page 63). The analysis undertaken in the reporting of the *Science Horizons* dialogue is insightful in this respect. It suggests that public views on future science and technology will be determined not simply on the benefits and risks of the particular technologies, but through the extent to which they respond to 'social goods', namely: better health (a social good); independence, especially for the elderly (a social good), convenience (a social good), quality of life (a social good), risks to safety (a social bad), scope for loss of privacy and autonomy (a social bad), social divisiveness (a social bad) and lack of genuine human interaction (a social bad). While each of the above points requires further differentiation and expansion, it nevertheless reinforces the observation that public views on science and technology depend critically on their 'social constitution', that is on the distinctive values and social assumptions that are embedded in their development (Grove-White *et al.*, 2000).

Ethics, trade-offs and inequality

A final theme concerns lay ethical judgement. Daniel Start (2010) in his review of the Sciencewise-ERC dialogues observes that differences of ethical opinion tended to be most pronounced within the dialogues rather than between them. A primary consideration was whether there was a sense of genuine social benefit from publicly funded science. At an individual level, where the social benefit was high, the public was prepared to accept higher trade-offs. Thus, in the *Stem Cell* dialogue, stem cell research was seen as acceptable only in cases where there existed the potential for very significant medical breakthroughs for the treatment of incurable diseases. In cases where stem cells were proposed in cosmetic applications or for the purposes of human enhancement, where the social benefit was seen as low, the research was seen as less acceptable. This kind of trade-off was commonplace in the *Honest Broker* dialogues.

A secondary consideration was the social distribution of those costs and benefits. Across many of the dialogues was a concern that the political economy of new science and technology would disproportionally impact upon vulnerable groups, particularly the poor, the ill, the unborn and those unable to defend themselves. Concern was expressed that nanotechnologies would benefit the rich and the powerful, not the poor or the unemployed (*Nanodialogues*); that medical research would be biased towards western and affluent illnesses rather than those in developing countries (*Stem Cells*); that the National DNA database could be used by governments to further discriminate against ethnic minorities (*Forensic Uses of DNA*); while the use of new drug treatments in the management of mental health conditions could be seen as a cheap alternative to social and behavioural therapy (*Drugsfutures*). This kind of consideration was again most common in the *Honest Broker* dialogues.

A final consideration was the differing and competing philosophical perspectives that people used to discuss the ethics of particular scientific and technological innovations. Start (2010) distinguishes two competing philosophies at work in the dialogues: a liberal and individualistic set of values and rights pitted against communitarian and collective values and virtues. Thus, while people were in general positive about the prospects of new technology to improve convenience, save time and add choice (appealing to liberal and individualistic values), at the same time they were wary that those same technologies would erode communities, devalue traditions and dehumanise relationships (*Science Horizons*). While there was strong concern about the perils of inappropriate drug use from a collective viewpoint, there was at the same time strong support for individuals to have the right to make their own decisions (*DrugFutures*). While there was strong support for the use of science and technology to aid national security, the public also spoke up for rights to privacy and anonymity (*Forensic Uses of DNA*).

Governance responses

To summarise, our review of the 17 public dialogues sponsored by Sciencewise-ERC highlights a range of concerns surrounding the governance of science and technology: whether research is being conducted in the public interest, whether the institutions overseeing research are to be trusted, whether people feel included or excluded in deciding what kinds of science gets funded, whether the speed of innovation processes exceeds their scope for regulatory and ethical oversight, and whether the risks and benefits of the research are unevenly distributed. In this section we now focus on the ways in which science and policy institutions are responding to these challenges, with examples taken from the fields of genomics, nanotechnology and climate science – three areas of high profile and ethically-sensitive science. The governance responses cover two main functions: as routes through which publics can have influence in shaping the nature and direction of emerging science and technology, and as mechanisms for public transparency, scrutiny and accountability.

Genomics

The institutional governance of genomics is influenced by public and political controversy of genomics-related technologies (such as GM crops), and attempts to regain public trust in the wake of the BSE crisis. This has facilitated a move away from a top-down centralised *government* regulatory approach, favoured in the 1970s and 1980s, towards a more distributed system of networked governance since the 1990s where state-led regulation coexists with more participative forms of policy-making (Gottweiss, 2005; Lyall, 2007). The latter is being promoted by principles of 'responsible governance', such as those espoused in the European Commission's Strategy for Life Science and Biotechnology (European Commission, 2007), which seeks to combine goals of ethical and social responsibility with science-based regulatory oversight.

In the UK this contributed to a number of institutional innovations, including the setting up of three independent arms-length advisory bodies, all with expertise and remits to consider the social and ethical implications of genomics technologies, with stated commitments to transparency and openness, and requirements for public input (Grove-White, 2001). The Human Genetics Commission was set up by the UK Government in 1999, comprising experts in genetics, ethics, law and consumer affairs, to provide advice on the social, ethical and legal issues associated with human genetics and its impacts on people and health care. The Food Standards Agency (FSA), was set up a year later to protect the public's health and consumer interests in relation to food, with parallel advisory responsibilities relating to GM foods. A further body, the Agriculture and Environment Biotechnology Commission (AEBC) was established in 2000 to provide strategic advice on biotechnology issues affecting agriculture and the environment until it was wound up in 2005.

The design of these bodies has involved a number of governance mechanisms to address issues of expertise, transparency, openness and public scrutiny, including: independence, a diverse membership, a commitment to openness and transparency, and mechanisms for embedded public scrutiny and representation. These bodies also were at the forefront of attempts to develop formal invited public dialogues as a way of eliciting public views and concerns over genomics developments. The AEBC led the *GM Nation?* national consultation on the commercialisation of GM crops in the UK in 2003 (Horlick-Jones *et al.*, 2007); the FSA held its own public dialogue processes on GM food in the same year to link with the Government's wider public debate on GM (FSA, 2003); while the HGC has run a number of consultations including the Sciencewise-ERC sponsored Citizens' Inquiry into the forensic use of genetic information in 2008 (as discussed earlier). These processes closely resembled the *Honest Broker Model* of public dialogue outlined above.

In contrast to this institutionally-sanctioned style of governance response are moves towards more distributed forms of innovation and public interaction with genomics research, including those of open innovation, crowdsourcing (a technique through which patient and online communities can contribute their information directly to large research datasets) and open source drug discovery (Prainsack and Wolinsky, 2010). The rationales and opportunities for such techniques are various: from merely extractive exercises aimed at efficiency and knowledge gains to wider ambitions of citizenship empowerment and the democraticisation of genomic knowledge (Knoppers, 2009; Gottweiss and Lauss, 2010).

'Uninvited' spaces of public engagement, typically organic and spontaneous processes orchestrated by citizens themselves rather than host institutions and involving forms of activism and protest, have long been associated with genomics-based science and technologies. Historically, these have tended to be dismissed and denied by science and policy institutions as either irrational and/or irresponsible. However, an alternative and more positive view of the potential for institutional reflexivity in this regard is possible, concerning the ways in which 'counter publics' can impact on the policy process (Warner, 2002). This has been shown by the Biotechnology and Biological Sciences Research Council (BBSRC) and other scientific actors in the UK where reflection on the reasons for public opposition to GMOs has meant that uninvited public engagement played a largely indirect role in reorientating UK plant and crop science research strategies from a narrow vision concentrated on GM to a more holistic, diverse and flexible portfolio that now includes non-GM approaches to crop improvement (see BBSRC, 2004; Doubleday and Wynne, in press).

Nanotechnology

The institutional debate on nanotechnology has been fuelled at least in part by a desire to learn lessons from the experience of GM crops in Europe, where arguably there had been a failure to recognise public concerns about the development of these technologies until after public resistance to their commercialisation had solidified (Kearnes *et al.*, 2006b). In this sense, nanotechnology has been represented as an opportunity to gain public input and explore social and ethical implications much earlier on in the innovation process, when it is still possible to shape the development of the emerging technology (Macnaghten *et al.*, 2005). Nanotechnology has also been viewed by social scientists and others as an opportunity to move the debate from a narrow focus on risk governance, where the questions are reduced to ones of risk and safety, to 'innovation governance' (Felt and Wynne, 2007), which emphasises 'upstream questions' of the sort routinely raised by publics in dialogues (see above), such as: 'Why this technology? Why not another? Who needs it? Who is controlling it? Who benefits from it? To what ends will it be directed?' (Wilsdon and Willis, 2004: 28). These questions have driven a set of initiatives aimed at the upstream engagement of nanotechnologies and the development of governance responses. The dominant response from science and policy institutions, at least in the early stages, has been to orchestrate managed spaces of small-scale public deliberation and citizen-scientist interaction, which adhere to the *Upstream Model* of public dialogue outlined above, to negotiate the social and ethical implications of emerging nanotechnologies (for UK examples, see Gavelin *et al.*, 2007; for a survey of European initiatives, see Stø *et al.*, 2010).

A further nanotechnology governance response, initiated to a large extent by the social scientific research community, has been the development of integrated systems of 'real-time technology assessment' (Guston and Sarewitz, 2002) and 'anticipatory governance' (Barben *et al.*, 2008), as demonstrated in the work of the Centre for Nanotechnology in Society at Arizona State University. Here forms of public engagement and dialogue such as those noted above, foresight practices, and reflexive collaboration between natural and social scientists are brought together in a comprehensive framework. This offers an integrated and systematic approach to building in continuous reflection on the social and ethical implications of nanotechnologies as they are being developed. It also highlights the importance of encouraging and building the capacity of nanoscientists in the laboratory to enact such reflection themselves with the help of, and in collaboration with, social scientists (Doubleday, 2007).

These largely discrete and contained experiments have led recently to institutional responses that are beginning to consider the wider governance system, and that seek to bring about the responsible development of nanotechnologies through more distributed and self-regulated means. This includes: (1) voluntary reporting schemes, such as the *Voluntary Reporting Scheme for Engineered Nanoscale Materials* developed by Defra as a mechanism for building evidence on possible risks; (2) voluntary codes of conduct and emerging mechanisms aimed at the responsible development of nanoscience and nanotechnologies, such as the European Commission's

Code of Conduct for Responsible Nanosciences and Nanotechnologies Research (European Commission, 2008); and (3) innovative experiments aimed at responsible innovation, such as the Engineering and Physical Sciences Research Council's (EPSRC) Nanotechnologies Grand Challenge for Environmental Solutions, which has trialled new anticipatory risk governance approaches in the form of risk registers aimed at identifying the wider potential impacts (social, environmental, ethical) of proposed research (Owen and Goldberg, 2010).

Climate science

Ever since the earliest stages in the formation of international action on climate change the framing of the debate has been dominated by climate science, which has assumed a linear relation to policy development (Pielke, 2010). This has shaped the governance of climate science, with appeals to scientific consensus seen as central to the policy goal of promoting action, and through a largely distant relationship between climate science and society, with interaction mainly occurring through the media. This seemingly cosy relationship between climate science, policy and society has been shaken over the past few years in the wake of the UEA hacked emails affair and questioning of the impartiality, accuracy and balance of IPCC scientific assessments. Recent events have led to increasing public scrutiny. Ongoing developments in the governance of climate science include: recommendations from the Muir Russell independent review into the UEA hacked emails emphasising the need for openness and transparency in relation to climate science, as well as for improvements in communication, peer review processes, and the handling of uncertainties (Russell *et al.*, 2010); and recommendations from the InterAcademy Council's (2010) recent review of the IPCC's governance and management, including its review process, its characterisation and communication of uncertainty, and its communications. Some of the most widespread and high profile responses have been initiatives to open up climate science data and codes that underpin climate change models to wider access by scientists and non-scientists (Kleiner, 2011).

However, there are questions as to whether this is a sufficient response. Is it enough simply to focus on communication, transparency and the opening up of data? Although constructive and necessary, such moves arguably leave the dominant framing of climate change, and the linear relationship between climate science and policy/action (as outlined above), untouched. An alternative approach has been developed by leading thinkers in *The Hartwell Paper: A new direction for climate policy after the crash of 2009* (Prins *et al.*, 2010). The paper questions the implied linear relationship between climate science and policy, which is critiqued as based on a 'flawed assumption that the solutions to climate change should be 'science driven' as if a shared understanding of science will lead to a political consensus' (Prins *et al.*, 2010: 17-18). *The Hartwell Paper* calls for a more humble and practical way of thinking about climate science that acknowledges the multiple framings inherent to debate on climate change (Hulme, 2009), and the role of value-judgements including their relation to science (Pielke, 2007), which need to be opened up to democratic deliberation. The group argue that the framing of the climate issue needs to be inverted: from a focus on sin to that of human dignity; from viewing climate change as a conventional tractable environmental problem to understanding it as a persistent condition that must be coped with; and from seeing 'climate policy' as a single, target driven, coherent and enforceable thing under which multiple issues reside to one where 'multiple framings and agendas are pursued in their own right, and according to their own logics and along their own appropriate paths' (Prins *et al.*, 2010: 10). In short, it is suggested that the restoration of trust in expert organisations depends on a radical reframing and reconfiguration of the relationship between climate science, climate policy and societal change.

Implications for critical risk research

We conclude by pointing to the implications of our review for critical risk research on technological governance. Our first conclusion concerns the imperative for research that engages with the cross-cutting themes as identified by lay publics. Understanding technological risk means to engage with the concerns that underpin and drive public response: the *purpose* of science and technology and the underlying *motivations* of scientists; the *trustworthiness* of responsible institutions and whether they are seen to act in the public interest; the scope for *inclusion* and the desire for publics to feed their *values* into the innovation process; the *speed* and *direction* of science and whether the pace of science is seen to exceed its scope for ethical and regulatory oversight; and whether the *culture* and *organisation* of science and innovation encourages or discourages reflection on risks, uncertainties and ethical considerations.

Our second conclusion stresses the need for research that recognises distinct trends in the ways in which science policy institutions are considering governance. These include the trend to move beyond formal governance processes – such as formal deliberative processes, ethical codes of conduct, ethical review, training and cultural change programmes – towards a more diverse range of ways in which institutions, scientists and publics can be exposed to governance issues. These range from 'uninvited' engagement spaces to various forms of outreach, voluntary codes of conduct, open sourcing, crowdsourcing and co-design. Research needs to recognise the more distributed forms of these processes, and to scrutinise the conditions under which they remain merely extractive exercises or whether they offer the public a genuine role in shaping the framing, direction and governance of scientific and technological developments.

Our third conclusion concerns the need for research to examine the efficacy of governance responses to the substantial governance concerns highlighted above. While some of these issues are at least partly responded to in governance practices (concerns about inclusion for example) others are not so evident (such as concerns over the purposes of emerging science and technology). For instance, in the case of nanotechnology, upstream questions relating to human needs and purposes are often reduced to ones of risk and impacts in actual governance practice. Thus, for example, while the trialing of the EPSRC's risk register in the Nanotechnology Grand Challenge was a significant innovation in governance, the emphasis on human health risk to researchers mitigated against upstream reflection of wider social implications.

Our fourth conclusion concerns the need for innovation in risk research and in particular for approaches that open up the processes of institutional responses that have all too often been 'black boxed' in past social science inquiry. Pointers to future research include: the need for more explicitly sociological and contextual approaches to help understand how governance responses take place in the context of a complex interplay of multiple actors, intermediaries and other influences; engagement with theories of organisational learning and change to help understand the possible barriers, drivers and influences of institutional response; and engagement with political economy approaches that seek to situate the governance of science and technology within wider contexts of economic and cultural globalisation, national economic competitiveness and corporate interests.

Our final remark concerns the requirement for scientific institutional cultures to reflect on their own cultures and their assumptions about others, including the public. In this sense the problem of public trust in science can be seen as 'a symptom of a continuing failure of scientific and policy institutions to place their own science-policy institutional cultures into the frame of dialogue' (Wynne, 2006: 211). The 'deeply-entrenched habitual tendency in science and governance to imagine possible learning as instrumental only' (Felt and Wynne, 2007: 18) – as has been shown to be the case in participatory governance of science and technology in the UK (Chilvers, 2010) – can limit the ability of institutions to fully understand and respond to the sorts of governance concerns highlighted above, which could be made possible only through more transformative, reflective and relational forms of learning.

References

Barben, D., Fisher, E., Selin, C. and Guston, D. (2008) Anticipatory governance of nanotechnology: foresight, engagement, and integration, in *The Handbook of Science and Technology Studies, Third Edition* (eds E. Hackett, O. Amsterdamska, M. Lynch and J. Wajcman), MIT Press, Cambridge MA.

Beck, U. (1992) *Risk Society: Towards a New Modernity*, Sage, London.

Beck, U. (2000) *World Risk Society*, Polity Press, Cambridge.

Beck, U. (2009) *World at Risk*, Polity Press, Cambridge.

Beck, U., Giddens, A. and Lash, S. (1994) *Reflexive Modernization: Politics, Tradition and Aesthetics in the Modern Social Order*. Polity Press, Cambridge.

Bingham, N. (2008) Slowing things down: Lessons from the GM controversy. *Geoforum*, **39**, 111–122.

BBSRC (2004) *Review of BBSRC-funded Research relevant to Crop Science*. http://www.bbsrc.ac.uk/web/FILES/Reviews/0404_crop_science.pdf

Chilvers, J. (2008) Environmental risk, uncertainty, and participation: mapping an emergent epistemic community. *Environment and Planning A*, **40**(12), 2990–3008.

Chilvers, J. (2010) *Sustainable Participation? Mapping out and reflecting on the field of public dialogue on science and technology*. Sciencewise Expert Resource Centre: Harwell.

Chilvers, J. and Macnaghten, P. (2011) *The Future of Science Governance: A review of public concerns, governance and institutional response*. Sciencewise Expert Resource Centre: Harwell.

Doubleday, R. (2007) Organizing accountability: co-production of technoscientific and social worlds in a nanoscience laboratory. *Area*, **39**, 166–175.

Doubleday, R. and Wynne, B. (in press) Despotism and democracy in the UK: experiments in reframing relations between the state, science and citizens, in *Reframing Rights: The constitutional implications of technological change* (ed S. Jasanoff), MIT Press: Cambridge, MA.

European Commission (2007) *Communication on the mid term review of the Strategy on Life Sciences and Biotechnology*, COM(2007) 175, European Commission, Brussels.

European Commission (2008) *A code of conduct for responsible nanosciences and nanotechnologies research*, European Commission, Brussels.

Felt, U. and Wynne, B. (2007) *Taking European Knowledge Seriously*. Report of the Expert Group on Science and Governance to the Science, Economy and Society Directorate. Directorate-General for Research, European Commission, Brussels.

Gavelin, K, Wilson, R. and Doubleday, R. (2007) *Democratic Technologies? The final report of the Nanotechnology Engagement Group (NEG)*, Involve, London.

Gottweis, H. (2005) Governing genomics in the 21st century: between risk and uncertainty. *New Genetics and Society*, **24**(2), 175–194.

Gottweis, H. and Lauss, G. (2010) Biobank governance in the post-genomic age. *Personalized Medicine*, **7**(2), 187–195.

Grove-White, R. (2001) New wine, old bottles? Personal reflections on the new biotechnology commissions. *The Political Quarterly*, **72**(4), 466–472.

Grove-White, R., Macnaghten, P., Mayer, S. and Wynne, B. (1997) *Uncertain World: genetically modified organisms, food and public attitudes in Britain*, Centre for the Study of Environmental Change, Lancaster.

Grove-White, R., Macnaghten, P. and Wynne, B. (2000) *Wising Up: The Public and New Technologies*, Centre for the Study of Environmental Change, Lancaster.

Guston, D. and Sarewitz, D (2002) Real-time technology assessment. *Technology in Society*, **24**, 93–109.

HEFCE (2007) HEFCE News 'Bridging the gap between higher education and the public.' See www.hefce.ac.uk/news/hefce/2007/beacons.asp

HM Treasury (2004) *Science and Innovation Investment Framework 2004–2014*, The Stationery Office, London.

Hoffmann-Reim, H. and Wynne, B. (2002) In risk analysis one has to admit ignorance. *Nature*, **416**(14 March 2002), 123.

Horlick-Jones, T., Walls, J., Rowe, G., Pidgeon, N., Poortinga, W., Murdock, G. and O'Riordan, T. (2007) *The GM Debate: Risk, Politics and Public Engagement*, Routledge, London.

House of Lords (2000) *Science and Society*, The Stationery Office, London.

Hulme, M. (2009) *Why We Disagree About Climate Change: Understanding Controversy, Inaction and Opportunity*, Cambridge University Press, Cambridge.

InterAcademy Council (2010) *Climate change assessments: Review of the processes and procedures of the IPCC*, InterAcademy Council, Amsterdam.

Irwin, A. (2006). The politics of talk: Coming to terms with the 'New' Scientific Governance. *Social Studies of Science*, **36**(2), 299–330.

Jasanoff, S. (2003) Technologies of Humility: Citizen Participation in Governing Science. *Minerva* **41**(3), 223–244.

Jasanoff, S. (2006) Technology as a site and object of politics, in *Oxford Handbook of Contextual Political Analysis* (eds C. Tilly, and R. Goodin), Oxford University Press, Oxford.

Kearnes, M., and Wynne, B. (2007) On nanotechnology and ambivalence: The politics of enthusiasm. *NanoEthics* **1**(2), 131–142.

Kearnes, M., Macnaghten, P. and Wilsdon, J. (2006a). *Governing at the Nanoscale: People, Policies and Emerging Technologies*, Demos, London.

Kearnes, M., Grove-White, R., Macnaghten, P., Wilsdon, J. and Wynne, B. (2006b) From Bio to Nano: Learning the Lessons, Interrogating the Comparison. *Science as Culture*, **15**(4), 291–307.

Kleiner, K. (2011) Data on demand. *Nature Climate Change*, **1**, 10–12.

Knoppers, B.M. (2009) Genomics and policymaking: from static models to complex systems? *Human Genetics*, **125**, 375–379.

Lezaun, J. and Soneryd, L. (2007) Consulting citizens: Technologies of elicitation and the mobility of publics. *Public Understanding of Science*, **16**(3), 279–297.

Lyall, C. (2007) Governing Genomics: New Governance Tools for New Technologies? *Technology Analysis and Strategic Management*, **19**(3), 369–386.

Macnaghten, P. (2010) Researching technoscientific concerns in the making: narrative structures, public responses and emerging nanotechnologies. *Environment and Planning A*, **41**, 23–37.

Macnaghten, P., Grove-White, R., Jacobs, M. and Wynne, B. (1995) *Public Perceptions and Sustainability in Lancashire: Indicators, institutions, participation*, CSEC, Lancaster.

Macnaghten, P., Kearnes, M. and Wynne, B. (2005) Nanotechnology, governance and public deliberation. What role for the social sciences? *Science Communication*, **27**(2), 268–291.

Macnaghten, P. and Urry, J. (1998) *Contested Natures*, Sage, London.

Owen, R. and Goldberg, N. (2010) Responsible innovation: a pilot study with the U.K. Engineering and Physical Sciences Research Council. *Risk Analysis*, **30**(11), 1699–1707.

Perrow, C. (1984) *Normal Accidents: Living with high risk technologies*, Basic Books, New York.

Pielke, R. (2007) *The Honest Broker: Making sense of science in policy and politics*, Cambridge University Press, Cambridge.

Pielke, R. (2010) *The Climate Fix: What scientists and politicians won't tell you about global warming*, Basic Books, New York.

Prainsack, B. and Wolinsky, H. (2010) Direct-to-consumer genome testing: opportunities for pharmacogenomics research? *Pharmacogenomics*, **11**(5), 651–655.

Prins, G., Galiana, I., Green, C., Grundmann, R,. Korhola, A., Laird, F., Nordhaus, T., Pielke Jnr, R., Rayner, S., Sarewitz, D., Shellenberger, M., Stehr, N., Tezuko, Hiroyuki (2010) *The Hartwell Paper: a new direction for climate policy after the crash of 2009*, Institute for Science, Innovation and Society, University of Oxford; LSE Mackinder Programme, London School of Economics and Political Science, London.

Ipsos MORI (2011) *Public Attitudes to Science 2011: Summary report*, Department for Business, Innovation, and Skills, London.

Renn, O. (2008) *Risk Governance: Coping with Uncertainty in a Complex World*, Earthscan, London.

Research Councils UK (2010) *Concordat for Engaging the Public with Research*. http://www.rcuk.ac.uk/per/Pages/Concordat.aspx

Royal Commission on Environmental Pollution (1998) *Royal Commission on Environmental Pollution 21st Report: Setting Environmental Standards*, The Stationery Office, London.

Russell, M., Boulton, G., Clarke, P., Eyton, D. and Norton, J. (2010) *The Independent Climate Change E-mails Review*, http://www.cce-review.org/pdf/FINAL%20REPORT.pdf

Science and Trust Expert Group (2010a) *Science for All: Report and action plan*. London: Department for Business, Innovation and Skills http://interactive.bis.gov.uk/scienceandsociety/site/all/files/2010/02/Science-for-All-Final-Report-WEB.pdf

Science and Trust Expert Group (2010b) *Starting a national conversation about good science*. http://interactive.bis.gov.uk/scienceandsociety/site/trust/files/2010/03/BIS-R9201-URN10-699-WEB.pdf

Start, D. (2010) *Ethical Dimensions in Sciencewise. A review of public perceptions of ethics issues from the Sciencewise dialogues*, Sciencewise Expert Resource Centre: Harwell.

Stengers, I. (2000) *The Invention of Modern Science*, University of Minnesota Press, Minneapolis.

Stirling, A. (1998) Risk at a Turning Point. *Journal of Risk Research* **1**(2), 97–110.

Stirling, A. (2005) Opening up or closing down? Analysis, participation and power in the social appraisal of technology, in *Science and Citizens: Globalization and the Challenge of Engagement* (eds M. Leach, I. Scoones, and B. Wynne), Zed, London.

Stirling, A. (2007) Deliberate futures: precaution and progress in social choice of sustainable technology. *Sustainable Development*, **15**(5), 286–295.

Stø, E., Scholl S., Jègou, F. and Strandbakken, P. (2010) The future of deliberative processes on nanotechnology, in *Understanding Public Debate on Nanotechnologies Options for Framing Public Policy* (eds R. von Schomberg, and S. Davies). European Commission, Brussels.

Warner, M. (2002) Publics and counter publics. *Public Culture*, **14**(1), 49–90.

Wilsdon, J. and Willis, R. (2004) *See-through Science: Why Public Engagement Needs to Move Upstream*, Demos, London.

Wynne, B. (1992) Misunderstood misunderstandings: social identities and the public uptake of science. *Public Understanding of Science*, **1**, 281–304.

Wynne, B. (2002) Risk and environment as legitimatory discourses of technology: Reexivity inside out? *Current Sociology*, **50**(3), 459–477.

Wynne, B. (2005) Risk as globalising 'democratic' discourse: Framing subjects and citizens, in *Science and Citizens: Globalization and the Challenge of Engagement* (eds M. Leach, I. Scoones, and B. Wynne), Zed, London.

Wynne, B. (2006) Public engagement as a means of restoring public trust in science–hitting the notes, but missing the music? *Community Genetics*, **9**, 211–220.

Appendix 1. Sciencewise-ERC Dialogue Projects (all available at http://www.sciencewise-erc.org.uk)

1 Animals containing human material (2010)
2 Big energy shift (2008–2009)
3 Community X-Change (2005–2008)
4 Drugsfutures (2006–2008)
5 Energy 2050 pathways (2010–2011)
6 Forensic use of DNA (2007–2008)
7 Geoengineering (2010)
8 Hybrids and chimeras (2006)
9 Industrial Biotechnology (2008)
10 Landscape and ecosystem futures (2011)
11 Low carbon communities challenge (2010–2011)
12 Nanodialogues (2005–2007)
13 Risky business (2005–2006)
14 Science Horizons (2006–2007)
15 Stem cells (2007–2008)
16 Synthetic Biology (2009–2010)
17 Trustguide (2005–2006)

CHAPTER 7

Technologies of Risk and Responsibility: Attesting to the Truth of Novel Things

Matthew B. Kearnes
School of History and Philosophy, University of New South Wales, Australia

> *In order to be responsible it is necessary to respond to or answer to what being responsible means.*
>
> (Derrida, 1995: 25)

Introduction

In recent years a range of low-probability but high consequence threats – the asymmetric threats of global terrorism, the uncertain implications of human-induced climate change and concerns about long term public health and population trends – have come to dominate the contemporary political imagination. In response risk management doctrine has been broadly redefined in preemptive rather than precautionary terms. In place of conceptions of risk as either predictable and calculable, risk management practice has been redefined to address these quiescent, but potentially catastrophic, harms. Mirroring these broader developments, in the specific field of technological risk governance, attention has begun to shift to the unanticipated consequences of new technologies. In contrast to approaches based on the 'precautionary principle' these preemptive strategies act on the potential for new technologies to herald novel risks and harms in the future. In order to address these unnamed threats preemptive and anticipatory forms of risk management are increasingly being incorporated into 'upstream' processes technological innovation, enabling a form of 'real time technology assessment' to augment the regulation of 'downstream' technological products (Barben *et al.*, 2008; Guston and Sarewitz, 2002; Wilsdon and Willis, 2004).

Critical Risk Research: Practices, Politics and Ethics, First Edition.
Edited by Matthew Kearnes, Francisco Klauser and Stuart Lane.

This preemptive turn represents a significant shift in the models of risk used in contemporary risk governance that were largely developed in the 1950s and 1960s as a response to the growth of the chemicals industry. The risks of these materials were conceptualised as a product of their innate toxicity and managed through the technical processes of risk assessment and the application tolerance limits, often set out in legislation. However, the development of new technologies, and particularly controversies around the release of genetically modified organisms, have forced a revision of these models, with debate centring on the uncertain and unanticipated combinational effects of new materials, released into the natural environment (Kearnes *et al.*, 2006). The recent development of a range of preemptive strategies in technological risk assessment – based on notions of 'anticipatory governance' and 'responsible development (Barben *et al.*, 2008; Owen and Goldberg, 2010) – therefore represent a response to this core problem: how to govern the development of new technologies while remaining sensitive to the uncertainties that these technologies entail. What we are witnessing is a significant shift in what Ewald (1999) terms, the 'paradigm of precaution'. Whilst the doctrine of precaution – which has been central to formal risk governance and particularly the 'precautionary principle' – depends on predictive and calculable assessments of possible threats, preemptive strategies entail speculative interventions 'over threats that have not yet emerged as determinate threats, and so does not only halt or stop from a position outside.... Preemptive acts become immersed in the conditions of emergence of a threat, ideally occurring before a threat has actually emerged' (Anderson 2010, 14).

Alongside this logic of preemption, technological risk governance is increasingly framed by notions of moral and ethical virtue and particularly conceptions of individual and corporate responsibility. Mirroring the broader *responsibilisation* of corporate behaviour (Shamir, 2008), contemporary regulatory strategies are characterised by initiatives that incentivise the 'ethically sensitive' development of technology (Franklin, 2001). Strathern (2005) argues that a set of overlapping temporalities are at work here, that in light of a range of anticipated but latent risks the future of technological development is represented as 'fragile'. It is argued that by investing scientific and technological development with notions of moral and ethical virtue will enable the 'responsible innovation' of new products. These strategies attempt to confer a form of social robustness on technological innovation, precisely by enabling regulatory institutions to address the moral virtues of scientific and commercial practice. Typically encoded in a range extra-legal regulatory forms – the use of 'codes of conduct' being particularly prevalent in recent risk management practice – notions of corporate virtue function to sustain preemptive approaches to risk assessment by recasting these practices in moral and ethical terms.

These two shifts in contemporary risk governance are indicative of the political constitution of formal risk assessment and the particular role that these regulatory techniques play in shaping institutional responses to the development of new technologies. In this chapter I take up an analytic approach to these developments by exploring the ways in which the mutually reinforcing discourses of preemption and moral responsibility are evident in strategies employed in the governance and regulation of nanomaterials. Representing a new class of materials, developed through the precise manipulation of nanoscale surface morphology, the development and commercialisation of nanomaterials has challenged accepted risk management paradigms. While the key commercial novelty of these products is their unique properties, in the absence of comprehensive research concerning the possible toxicity and combinational effects of these materials, the potential risks of these materials subsist in their latent *futurity*. These risks escape and challenge accepted methods for characterising predictable risk pathways.

As regulators have begun to develop anticipatory mechanisms for managing these potential risks, in the following sections of this chapter I attend to what, Diprose *et al.* (2008) term the 'moral and economic underside to approach[es] to risk' (p. 269). What I mean by this is not simply that risks, and institutionalised practices of risk management, have a 'moral side' or are of significant ethical import. Rather in this chapter what I am interested in detailing are the ways in which questions of ethics, morality and virtue are called upon to sustain anticipatory and preemptive political logics and rationalities. Developing Foucault's genealogy of ethics – and particularly his analysis of the 'moralisation of the market' – in the following sections of this chapter I trace the development of these twined logics of preemption and responsibility through an analysis of strategies deigned to incentivise the 'responsible development' of nanomaterials (Royal Commission on Environmental Pollution, 2008). I argue that while these strategies are indicative of a wider politics of responsibility – characterised by conspicuous displays of corporate virtue (Shamir, 2008) – the mutually reinforcing logics of preemption and responsibility function to render the novel and hybrid materials of advanced technology both culturally readable and politically tractable.

The virtues of preemption

How, then, to trace the intersections between the often technical processes of formal risk assessment and the increasingly moralistic tone of contemporary risk governance, infused as they are with notions of corporate and personal responsibility? In a review of contemporary risk research, O'Malley (2000) outlines the conceptual problem here, suggesting that 'there is nothing obvious about how truth, morality and risk are now related to each other, for this has a tortured genealogy' (p. 457). O'Malley's genealogical approach, drawing on the work of Michel Foucault, stands in contrast to the dominant ways in which questions of ethics and morality have been presented in contemporary risk research. Given the influence ethical principalism and behaviourist theories in the social sciences more generally, the ethical dimensions of risk research are typically portrayed as classical dilemmas requiring trade-offs between contending ethical positions. For O'Malley this principalist approach captures but a fraction of the wider set of exchanges between risk, ethics and morality, suggesting that this approach separates questions concerning the 'ethical dimensions' of risk from the ways in which ethics and morality are put to use in designating potential threats as risks and in framing culturally acceptable responses to these dangers.

In their well known work on 'risk selection' – an analysis of the ways in which cultures 'decide what risks to take and which to ignore' (p. 1) – Mary Douglas and her collaborator Aaron Wildavsky (1982) make a similar point. They suggest that the ways that potential risks and harms are articulated is underscored by appeals to threatened moral orders and, further, that the culturally sanctioned responses to risk function to reinforce and re-establish these orders. The link between questions of morality and risk is, for Douglas and Wildavsky, a structural expression of the topologies of cultural and political power (see also Ericson and Doyle, 2003; O'Malley, 2004). Tracing attempts to define some risks in moral terms (for example gambling and drug taking) and to represent other potential risks (such as financial speculation and insurance) as virtuous, O'Malley (2000) suggests that in each of these struggles 'different models of risk were deployed, with parallel moral issues and various truths being mobilized' (p. 457).

What these insights suggest is that the differential processes by which risk and threats are articulated is necessarily tied to notions of morality and virtue. Institutionalised processes of risk assessment and governance might therefore be understood as part of a wider set of cultural and political struggles to define the 'moral truth' of predicted threats – and particularly of the risky and hybrid materials produced through technological innovation. The more recent invocations of corporate virtue and personal responsibility – what Diprose *et al.* (2008) term a 'paradigm of prudence'; a wary and anxious approach to the potential of future threats that 'urges societies [to] be in a constant state of readiness' (p. 269) – might therefore be understood as an expression of this mutually reinforcing relationship between risk and contemporary morality. Notions of virtue and responsibility are called upon in constitution of a set of political rationalities about need to intervene in the present so as to regulate an unpredictable future-to-come (Diprose, 2006). Whilst the latent potentiality of high-consequence/low probability threats form the backdrop to recent preemptive risk management strategies, attempts to engender new forms of personal and corporate responsibility constitute the ethical underbelly of these strategies, deployed in order to substantiate the moral fidelity of this preemptive rationale.

In composing what he terms a 'genealogy of ethics' Foucault (1994) develops an analytic approach that resonates with the institutional functionality of ethics and morality. Across a series of his later works Foucault (1990; 2008) turns to an analysis of the classical ethical systems of Hellenist Greece and the more recent invocations of ethical norms in contemporary liberalism. In each case Foucault 'seeks to identify the elements – techniques, subjects, norms – through which the question of 'how to live' is posed' (Lakoff and Collier, 2004: 420). For Foucault questions of ethics and virtue are not simply abstract principles. Rather he explores the ways in which ethical systems are constitutive of contemporary political rationalities and subjectivities. For example, in his analysis of the constitution of contemporary liberalism, defined by notions of individual freedom and autonomy, Foucault identifies an emerging moral topography based on an ethics of 'self cultivation' and 'care for the self'. For Foucault, ethics do a kind of work – they are put to work in constructing and maintaining certain kinds of social and political order and defining modes of subjectivity (Burchell, 1993).

Whilst the breadth of Foucault's diagnostic – and its intersections with the contemporary political rationalities – is beyond the scope of this chapter, what is significant about Foucault's account of ethic is the way he locates contemporary moral discourse its historic political and economic context. For Foucault, the emergence of political liberalism is enabled by the construction of a new political subjectivity, that of the free individual who becomes a loci of inalienable political and economic rights. In liberal thought it is the economic subject – the *homo œconomicus* – that becomes 'the surface of contact between the individual and the power exercised on him' (Foucault, 2008: 252). For Foucault earlier notions of self cultivation, evident in Protestant theology, and moral virtue are transformed with the advent of political liberalism, as the market assumes a preeminent role in arbitrating questions of moral value. In liberal thought the market is moralised – economic activity is defined as a morally virtuous pursuit whilst the market itself becomes a site for determining questions of moral worth and value. It is in this light that Foucault argues in the eighteenth century 'the market no longer appeared as a site of jurisdiction'. He argues rather that:

> Inasmuch as it enables production, need, supply, demand, value and price, etcetera, to be linked together through exchange, the market constitutes a site of veridiction, I mean a *site of verification*-falsification for governmental practice. Consequently, the market determines that good government is no longer simply government

that functions according to justice. The market determines that a good government is no longer quite simply one that is just. The market now means that to be good government, government has to function according to truth. (p. 31-32, emphasis added).

Foucualt's characterisation of the interplay between ethics, morality and liberal political and economic theory has the effect of reframing ethical traditions of self-cultivation in economic terms. The market itself becomes a moral object and a site for the arbitration – indeed *veridiction* – of moral truths. Shamir (2008; 2010) develops Foucault's notion of the moralisation of the market by showing how contemporary corporate capitalism has been *responsibilised*. What Sahmir means by this is that over the last 30 years, and particularly with the development of programmes of corporate social responsibility, business ethics and assessments of corporate virtue have become the stuff of capitalist exchange. For Shamir, virtue has become a commodity, traded between corporate actors in conspicuous public performances (see also Vogel, 2005). In his review of the development, and extensive use of distributed forms of corporate self-regulation, Shamir positions the rise of corporate social responsibility as 'a product of a capitalist crisis of legitimacy that occurred precisely at a moment when multinational corporations achieved unprecedented economic and political powers'. Programmes of corporate self governance and social responsibility are indicative of a 'the capitalist response to this crisis [and] an effort to invest corporate entities with a moral capacity that would in turn justify greater reliance on socially responsible private and self regulation' (2010, p. 533).

Similar dynamics are at play in the ethicisation of contemporary technological governance (Gottweis, 1998). Indeed the political factors that have inspired the development of these regulatory initiatives broadly parallel Shamir's analysis of the growth of business ethics and programmes of corporate social responsibility. In light of a broad crisis in the political legitimacy of risk governance, induced by a series of technological risk controversies surrounding, 'grammars of responsibility' have began to gain traction in official risk management doctrine.[1] For example, the UK government enquiry into the crisis surrounding the outbreak of Bovine spongiform encephalopathy (BSE) and the unanticipated onset of Creutzfeldt–Jakob disease (vCJD) identified a series of institutional failures in processes of risk assessment and communication, principally the tendency to overstate levels of scientific certainty 'shaped by a consuming fear of provoking an irrational public scare' (Phillips *et al.*, 2000: 264). The

---

[1] I am drawing here on Barnett *et al.*'s (2010) notion of 'grammars of responsibility'.

enquiry argued that future risk assessment and communication strategies need to be guided by the norms of openness and transparency, so as to restore public trust in the credibility of regulatory institutions. Similarly, in response to controversies relating to the development and release of genetically modified food, and the more recent debates about the governance and regulation of nanotechnology, regulators have begun to emphasise the need to engage in anticipatory risk assessment, to identify areas of potential risk before they are realised in commercial products (Royal Commission on Environmental Pollution, 2008; Royal Society and Royal Academy of Engineering, 2004). Mirroring Shamir's responsibilisation thesis, the response to these risk controversies has been to endow regulatory authorities with moral capacity for openness, responsibility and virtue in the anticipatory governance of new technologies.

For Shamir, while business ethics and notions of virtue operate as the underlying terms of discursive contemporary capitalist strategy, programmes for corporate social responsibility function as a social technology for conferring corporations with a capacity for moral action and conscience. Following Foucault's geneaology of the moral topology of political liberalism we might characterise technological risk governance in similar terms, as a technology of *veridiction*. Given that contemporary risk governance has tended to function as an institutionally sanctioned arbiter of the moral meanings of new technologies, the ways in which regulatory institutions, corporations and individual scientists are invested with a moral capacity is indicative of an attempt to resuscitate the social and cultural legitimacy of these institutional arrangements. Given this diagnosis the question that remains is how notions of virtue, morality and ethics sustain an anticipatory and preemptive approach to the incalculable threats of novel materials. In order to understand this mutually reinforcing relationship between the logics of preemption and those of responsibility I first consider the moral topologies of precautionary approaches to risk governance – noting how the epistemological constitution of the precautionary principle is indicative of a set of predictable and calculable threats – and then detail how the development of a range of novel and functional materials challenges this approach. In considering discourses of corporate virtue, evident in a range of regulatory initiatives for the responsible development of nanomaterials, I suggest that these mutually reinforcing political logics function to substantiate the moral truth of these hybrid materials.

Risk governance and the truth of things

Precautionary approaches to technological risk assessment are indicative of what Latour (2004) characterises as an 'impossible distinction between the *truth of things* and the will of humans' (p. 148, emphasis added). That is to say that the epistemological constitution of traditional risk assessment invokes a particular kind of moral topology, where risks are understood as an innate quality of risky materials. Typically entangled in political arguments that seek to distinguish matters of scientific fact from those of cultural judgement, contemporary risk governance has tended to conceptualise risk as a function of probabilistic exposure to risky materials exaggerated by 'human error' or regulatory oversight. In these models risk is understood as 'originat[ing] in the inanimate world, although human behaviour can exacerbate its intensity' (Jasanoff, 1999: 142). Informed by linear and mechanistic notions of causation, the axiomatic goal of contemporary risk governance has tended to be understood as an attempt to quantify and measure 'zones of technical uncertainty' in order to make risky materials governable.

This materialist ontology is tied to what we might think of as the 'operative logic' of formal risk assessment. Ewald (1999) characterises this logic as a 'paradigm of precaution' where the materialist conception of risk as embodied in risky substances invokes a set of institutionalised practices aimed at identifying, measuring and preventing harm. He suggests that:

> The notion of prevention . . . presupposes and accompanies the promotion of the notion of risk and, which comes down to the same thing, of measurable risk. Prevention presupposes science, technical control, the idea of possible understanding and objective measurement of risks. (p. 58)

In areas as diverse as the environmental damage caused by chemical pollution (Brickman *et al.*, 1985), the disposal of nuclear waste (Wynne, 1982) and most recently the development and release of genetically modified organisms (Jasanoff, 2005) – the logics of precaution and prevention have become constitutional in the assessment and management of possible threats. The 'precautionary principle' outlined – in the 1992 *Rio Declaration* and later adopted as a formal part of European risk regulation – encapsulates this logic. Suggesting that 'where there are threats of serious or irreversible damage, lack of full scientific certainty shall not be used as a reason for postponing cost-effective measures to prevent environmental degradation (United Nation, 1992: Principle 15), the principle encodes systems of preventative risk management on the basis of identifiable and calculable threats – even where the full extent of these possible threats remain uncertain.

In this model, risks are a product of material properties – rather than being indicative of the kind of structural link between the production of risk and technological modernity that Beck (2004) suggests is characteristic of a 'world risk society'. It is in this light that Jasanoff (1999) conceptualises institutionalised practices of risk assessment as performing a specific cultural function as a 'songline' of contemporary risk society's anxiety about its own technological achievements' (p. 141) precisely by rendering risk as a set of governable and measurable materialities, divorced from wider questions of human intention and cultural value. Precautionary approaches to risk assessment imply a mutually reinforcing relationship between the technical logics of prevention and prediction and the moral virtues of precaution and care. Where uncertainties subsist in critical risk research precautionary approaches imply that this will subside simply with more extensive risk research, building a more complete characterisation of threats and harms. The moral virtues of precaution and care are therefore called upon to sustain a particular materialist ontology that divorces the assessment of risk from wider questions concerning the pace and direction of technological change. In practice the precautionary principle sets up an institutionalised trade-off between risk assessment, technical uncertainty and 'cost-effective' remedial actions; as such it forms part of the broader set of political and economic technologies invoked in making technological products commercialisable. Logics of precaution, calculation and prediction therefore function to enable commercial innovation by suggesting that unless risks can be predicted and calculated in advance technological development should proceed, unhindered by wider questions concerning the social and cultural meanings of new technologies (Levidow, 2001).

Hybrid materialities

As outlined by Macnaghten and Chilvers (this volume), a series of recent technological risk controversies have challenged the continuing reliance on the precautionary principle. Most significantly these controversies have served to problematise the ways in which uncertainty is conceptualised. In debates concerning the risks of genetically modified food, the disposal of nuclear waste and the development of novel nanomaterials, the persistence of scientific uncertainty has been revealed as a constitutive condition of risk management. Rather than reducing uncertainty, risk research tends to expose further complexity, particularly in relation to the combinational effects induced by the release of novel materials in human bodies or to the natural environment (Sarewitz *et al.*, 2000).

The key issue here is the degree to which the virtues of precaution and the logics of prevention are tied to a materialist ontology that divorces the risks of nonhuman materials from questions of human agency and intention. The fragility of this ontology is especially poignant in the case of functional and novel nanomaterials. The development of nanomaterials is enabled by a capacity to manipulate the materials at the nanoscale, and inspired by a set of informational metaphors that emerged in scientific and technological practice after the Second World War. After the developments in control theory and information science understandings of the physical world in informational terms gained currency in areas of research such as molecular biology, systems theory and cognitive science (Dupuy, 2000; Keller, 1995). Much like the ways in which the genetic research is framed by the figure of DNA as a 'rule book' or 'biological script', research in nanotechnology is influenced by a conception of the physical world as 'informational'. Barry (2005) characterises the kind of novel materials that are produced through this kind of research as 'informationally enriched', produced through a technoscientific orientation 'concerned with the invention of... informed materials' (p. 52). In nanotechnology this informational enrichment operates in two ways. Firstly, a reductionist logic has the effect of conceptualising the material world as fundamentally malleable – that it can be manipulated *as if it were* information. Secondly, research in nanotechnology is inspired by the goal of producing materials with a degree of functionality, materials enriched with a degree of informational agency. However, it is also this informational enrichment that presents a challenge for contemporary risk management. The potential risks that these materials pose are not simply a product of their innate materiality but is rather induced by their hybrid functionality. For example, a recent study conducted by the Royal Commission for Environmental Pollution (2008) on the governance and regulation of nanomaterials articulated the regulatory challenge posed by such materials in the following terms:

> The key factors that should drive our interest in the environmental and human health issues surrounding novel materials are, indeed, their functionality and behaviour... [rather than]... the particle size or mode of production of a material. (p. 4)

That ordinarily stable materials are enriched with a set of novel functions – a hybrid mixture of material properties and human intentions – challenges the ontological constitution of formal risk assessment. Though nanomaterials are increasingly being incorporated into range of commercial, domestic and industrial products – including cosmetics, surface preparations and mechanical components[2] – concerns remain about the possible risks to both human health and the environment arising from the use of these materials. However, much like the controversies associated with the development of asbestos and thalidomide, the risks of nanomaterials

---

[2] For an inventory of products that incorporate the use of nanoscale materials see: www.nanotechproject.org/inventories/consumer/.

remain largely latent. The potential risks of nanomaterials are only likely to be substantiated after sustained use in commercial products and consequently after high levels of public exposure. Though a number of possible risk pathways have been identified – together with early indications of the potential toxicity of some engineered nanomaterials (Defra, 2007) – the potential for unpredicted risks constitutes the core political challenge posed in the governance of regulation of these materials. This challenge is made more pressing by both the morphological novelty and functionality of these materials. In a similar review of the possible risks posed by nanomaterials, Fiorino (2010) outlines the nature of this challenge:

> Nanotechnologies present complex and distinctive challenges to the public and private institutions responsible for managing environmental and health risks in society. In this sense, they are characteristic of many new environmental issues. Risks that may be associated with them are difficult to assess. They represent a rapidly growing economic sector and constantly evolve with changes in technology, markets and products. They do not appear to fit neatly into any set of legal frameworks that have been developed in the past or exist today. In addition, more so than some of the earlier generations of environmental issues, nanotechnologies in themselves offer potentially huge environmental and health benefits, along with the possibility of new and novel risks. In many respects, the issues associated with nanotechnology are more typical of the future of environmental problem-solving than those of large manufacturing sources and high-volume commodity chemicals that determined the design and application of environmental statutes in the past four decades. (p. 8)

Here the problem of technical uncertainty is given a new twist. Rather than representing simply a *lack* of knowledge – concerning the possible toxicity of new materials – these studies suggests that the regulatory challenge posed by novel materials is their capacity to function in unpredictable ways.

Preemptive governance of nanomaterials

In order to attend to this latent potential, the risk management practices employed in the governance and regulation of new materials have taken on a distinctive 'anticipatory turn'. After outlining the conceptual challenge posed by the governance of functional materials, the Royal Commission on Environmental Pollution (2008) study concluded by arguing for a preemptive and adaptive strategy:

> Novel materials, like other emerging areas of technology, require an adaptive governance regime capable of monitoring technologies and materials as they are developed and incorporated into processes and products. An effective, adaptive

> governance regime will have to be capable of applying the indicators of technological inflexibility identified in the technology control dilemma to decide when to intervene selectively in areas where it deems that a material represents a danger to the environment or human health. (p. 8)

For the Royal Commission the challenge posed by the development of functional nanomaterials suggests that risk management procedures need to be incorporated into the 'upstream' development of the technologies, rather than being simply confined to the 'downstream' assessment of products. This innovation in risk assessment is structured by a preemptive logic where threats are acted upon before their formal identification, through the development of adaptive and flexible forms of innovation governance. This logic of anticipation is evident in a range of overlapping initiatives: the use of foresight and early warning mechanisms to build anticipatory intelligence on possible risks, the proliferation of modes of corporate self-regulation and the development of programmes of public engagement and public deliberation in nanotechnology research programmes (Kearnes and Rip, 2009).

These preemptive strategies therefore mark a distinct shift in the operative logics of contemporary risk governance. Though premised on the need to act in conditions of scientific uncertainty, the logics of prevention and precaution 'operate in an objectively knowable world in which uncertainty is a function of a lack of information, and in which events run a predictable, linear course from cause to effect' (Massumi, 2007: 5). In this sense, the logics of precaution and prevention are 'separate from the processes [they] act on... [and] begin once a determinate threat has been identified, even if that threat is scientifically uncertain' (Anderson, 2010: 13). In comparison, notions of preemption are defined by an alternative and anticipatory temporality. The development of anticipatory and preemptive modes of risk management in the specific area of technological risk and novel materials is indicative of the degree to which a range of 'total threats' – associated with terrorism and climate change, for example – have come to dominate the political imaginary (de Goede and Randalls, 2009). The potentiality of these vague, yet catastrophic, risks has in turn created the conditions for an 'intensification of efforts directed toward [the] management of incalculable risks of life-threatening events of low probability but potentially catastrophic consequence' (Diprose *et al.*, 2008: 268). In this sense the preemptive strategies adopted to address the potential threats of novel materials shift both the technical calculus and loci of intervention. Rather than being premised on the curative actions to combat objectively describable threats political logics of preemption justify anticipatory intervention in the future inherent in the present actions.

This logic of preemption is a speculative one – justifying interventions against potential threats *as if they were* objectively defined risks. In the case of novel materials, the key issue concerns the latency of the potential risks – that existing materials, currently deemed safe, may reveal unanticipated risks in later years. For example, UK regulatory responses to the development of novel materials have been characterised by an attempt to generate early intelligence on the use of nanomaterials in existing commercial products and processes, by building an evidence base on the extent of nanomaterial use through a *Voluntary Reporting Scheme for Engineered Nanoscale Materials* (Defra, 2008). In interview a representative of the regulatory body, the UK Department for Environment, Food and Rural Affairs (Defra) outlined the regulatory strategy adopted in relation to nanomaterials:

> Without regulation in place, we are relying on industry to do the right thing. But we don't really know how they're doing the right thing. . . . So there's a twin track you need to ensure the responsible development of nanotechnologies while ensuring that risks are mitigated whenever possible. (Interview with representative of Defra, 24 September 2009)

The approach taken by the European Union is characterised by a similar kind of initiatives – programmes for the coordination of European research efforts, reviews of existing legislation and the development of the *European Code of Conduct For Responsible Nanosciences and Nanotechnologies Research* (European Commission, 2009). In a review of these initiatives von Schomberg (2010) outlines both the regulatory challenge posed by nanomaterials and the response taken by the European Commission. He suggests that:

> While, in the absence of a clear consensus on definitions, the preparation of new nanospecific measures will be difficult, and although there continues to be significant scientific uncertainty on the nature of the risks involved, good governance will have to go beyond policy making focused on legislative action. Schomberg (p. 8)

In response von Schomberg suggests that:

> The European Commission has responded to this with its adoption of a European strategy and action plan on nanotechnologies, which addresses topics from research needs to regulatory responses and ethical issues to the need for international dialogue. This strategy above all emphasizes the "safe, integrated and responsible" development of nanosciences and nanotechnologies. (p. 8)

In both cases these regulatory strategies reflect a concern for the unnamed and incalculable threats posed by nanotechnology. The risks of nanomaterials subsist in their potentiality, rather than as clearly identified threats. The widespread use of voluntary measures represents an attempt to generate some degree of preemptive traction over these latent risks. In practical terms the virtues of responsibility also operate to stimulate innovation in nanotechnology whilst also ensuring that adaptive and anticipatory structures are in place to deal with potential risk management issues and more substantial social and ethical questions. Significantly, notions of the responsible development of nanotechnology operate as a distillation of the problem inherent in conceptualising the governance of nanotechnology – that is: how to govern in such a way that enables further development and commercialisation of nanotechnology products under the conditions of uncertainty. In the following sections of this chapter I consider a range of recent initiatives on the 'responsible development' of nanotechnology and consider how notions of corporate virtue are presented as commercially strategic.

The value of virtue

The emergence of doctrines of preemptive risk management might be understood as an attempt to cultivate a set of social and institutional capacities that 'bridge the cognitive gap between present and future' where:

> anticipatory governance implies that effective action is based on more than sound analytical capacities and relevant empirical knowledge: It also emerges out of a distributed collection of social and epistemological capacities, including collective self-criticism, imagination, and the disposition to learn from trial and error . . . [T]he concept of 'anticipation' is meant to indicate, the co-evolution of science and society is distinct from the notion of predictive certainty. In addition, the anticipatory approach is distinct from: the more reactionary and retrospective activities that follow the production of knowledge-based innovations-rather than emerge with them. (Barben, *et al.*, 2008: 992)

What is striking about this notion of anticipatory preemption is the way in which it is conceptualised as the cultivation of an ethos. Whereas the paradigm of precaution is based on a prescriptive, if contested, logos – a legal doctrine ratified in contemporary risk management practice – here the logic of preemption is conceived in an altogether more distributed fashion. The aim of anticipatory approaches is to foster a moral sensibility for the future-to-come and the risks latent in novel materials. For the architects of these initiatives this is a sensibility that is located in scientific practice itself and in the political structures that that support the development of new areas of research.

The pathos involved in public performances of corporate responsibility replaces the altogether stricter legal morality of accountability and compensation. As introduced above Shamir (2004; 2008) characterises the increasing prevalence of notions of corporate virtue as 'an age of responsibilisation' conceptualising corporate social responsibility as a commercial strategy for the conspicuous performance of public virtues. In their analysis of the governance of biotechnology Brown and Michael (2002) come to a similar conclusion, noting that regulatory strategy in this field is increasingly defined by the performance of a form of 'corporate suffering'. In a political context where notions of authenticity have replaced those of regulatory authority, contemporary risk governance is increasingly characterised by strategies that perform a kind of openness and transparency as a way of rebuilding institutional credibility. What is striking about strategies designed to incentivise the responsible development of nanotechnology is the degree to which 'grammars of responsibility' are deployed to indicate a capacity for regulatory institutions to incorporate and respond to political critique. For example, in the 'business case' presented for the development of forms of corporate self-regulation, economic and commercial conceptions of the 'value of virtue' are paramount. Scientific institutions are endowed with a capacity for virtuous conduct and self-regulation as a way of guaranteeing the cultural credibility of these institutions (Brown and Michael 2002; Gottweis, 2008; Wynne, 2006). For example, in a briefing paper for the proposed UK *Responsible NanoCode*, developed through a partnership between the Nanotechnology Industries Association and the Royal Society, Sutcliffe and Hidgson (2006) outline the business case for the code in the following way:

> The challenge for business, therefore, is whether its technology development and commercialisation process is sufficiently inclusive to understand and mitigate risks from these wider uncertainties. There is an essential need for good quality transparent research into the environmental and health risks, and it may be in the long-term interests of business to play a role in filling this gap. Business must convince investors, insurers, NGOs, government, the media and perhaps most importantly the general public that it understands the technology and is taking a responsible approach, which will require a very open style. There is also a case that business has a role in helping get the right legislative and commercial framework to allow it to bring the technology to market safely and profitably. (p. 16)

In this case, the development of a voluntary code is presented in strategic terms as an initiative that will assist commercial interests in 'helping to get the *right* legislative and commercial framework'. The subtext here is that the code is a response to a strategic uncertainty – that of the anticipated state-led regulation on nanotechnologies. This double move is a characteristic feature of corporate self-regulation and voluntary codes that enable corporations to publically demonstrate and perform best practice whilst anticipating the possibility of more formal, government-led regulation. The promoters of the *Responsible NanoCode* seek to make a case for broader commercial interest in the code on the basis that subscription to the code will enable to corporations a credible basis upon which to influence subsequent regulatory policy. The *Code of Conduct: Nanotechnology* developed by BASF is presented in similar strategic terms:

> Along with offering opportunities, all new technologies also pose risks and this is true for nanotechnology, too. In order to tap into the opportunities offered by technological progress, we want to use new technologies when manufacturing innovative and market-grade products. Only on the basis of these concrete products can a rational assessment be conducted of the potential risks, compared with the opportunities, these products pose. This means that only the willingness to pursue opportunities and risks on a gradual basis will make innovations based on new technologies possible. (BASF, no date, 1)

BASF's intention here is to present its strategy as providing the basis for 'rational assessment' of potential risks – anticipating the possibility for the development of more stringent regulations on nanotechnology, and positioning BASF accordingly. The code therefore operates in an anticipatory fashion, as a demonstration of best-practice. By developing an information base on the use of nanomaterials the code functions to enable BASF to stake out a strategic role in shaping anticipated future regulations. Similarly, the *Nano Risk Framework,* developed through a collaboration between EDF and DuPont code situates itself as a 'systematic and disciplined process to evaluate and address the potential risks of nanoscale materials' (DuPont and Environmental Defense, 2007: 7). The Framework presents itself as an 'information led' approach that develops a rational basis for future regulation. It is through this 'information-led' process that the authors of the framework suggest that 'the adoption of this Framework . . . support the development of a practical model for reasonable government policy on nanotechnology safety' (DuPont and Environmental Defense, 2007: 7).

The ambition of these initiatives is therefore to provide a mechanism for the public performance of corporate virtue and responsibility. Though these initiatives draw inspiration from discourses of corporate social responsibility – and particularly the *Responsible Care Programme* developed in the late 1980s in the chemicals sector – when viewed as a site and technology of *veridiction* such initiatives also speak of a wider and more enduring set of cultural meanings. The overlapping logics of preemption and responsibility are indicative of the degree of economic and cultural investment in the informational enrichment of novel materials. Such materials are typically presented as heralding an unprecedented set of commercial opportunities. In this sense the twin logics of preemption and responsibility represent a cultural strategy that seeks to render such materials commercialisable precisely by attending to this latent hybridity. In a political context defined by notions of authentic transparency and corporate virtue, by enabling the public performance of a form of anticipatory responsibility these initiatives function in two ways. They serve to resuscitate the contested legitimacy of process of formal risk assessment and regulatory oversight while marking out the hybrid materialities of nanotechnology as both governed and governable.

Conclusion

In thc opcning chapter of his major work *After Virtue* the British philosopher Alasdair MacIntyre (1999) invites his readers to imagine 'a disquieting suggestion'. He depicts a fictional world where the natural sciences have 'suffer[ed] the effects of a catastrophe' where 'a series of environmental disasters are blamed by the general public on the scientists' and in the resulting political fallout 'widespread riots occur, laboratories are burnt down, physicists are lynched, books and instruments are destroyed' (p. 1). In this fictional scenario, where science and education have largely been destroyed, MacIntyre imagines the possibilities for reviving scientific learning. He speculates on how the collective achievements of physics, chemistry and biology be pieced together from the scraps of knowledge that survive their destruction. In this scenario all that the citizens of this imagined place possess, in order to accomplish this revival, are 'fragments: a knowledge of experiments detached from any knowledge of the theoretical context which gave them significance; parts of theories unrelated to the other bits and pieces of theory'. In this context MacIntyre suggests that these revived forms of physics, chemistry and biology, pieced together from assorted pieces of earlier knowledge, would bear little relation to the natural sciences, owing to the fact that the 'contexts which would be needed to make sense of what they are doing have been lost, perhaps irretrievably' (p. 1).

Of course for MacIntyre this story serves as an allegory for the revival of questions of morality and virtue in contemporary social and political life. In light the revival of questions of virtue in public life, MacIntyre's central argument is that 'contemporary ethical discourse lacks such a stable cosmos or a teleological understanding of human nature to guide ethical reasoning' (Lakoff and Collier, 2004: 422). He argues:

> What we possess are the fragments of a conceptual scheme, parts which now lack those contexts from which their significance derived. We possess indeed a simulacrum of morality, we continue to use many of the key expressions. But we have – very largely, if not entirely – lost our comprehension, both theoretical and practical, of morality. (p. 2)

This broad-ranging argument is at once philosophical, anthropological and historical. MacIntyre suggests that though questions of virtue and morality are increasing preoccupations in contemporary public life the twin processes of secularisation and globalisation have rendered any attempt to pose universal conceptions of human ends fundamentally problematic. Contrasting contemporary social life with that of classical Greece, MacIntyre notes both the impossibility of sustaining a common understanding of human nature and the incoherence of contemporary ethical discourse. The proliferation of questions of ethics and responsibility is for MacIntyre symptomatic of the fact that 'morality *has* become *generally available*' becoming a 'mask for almost any face' (p. 110, emphasis in original).

Viewed from the perspective of strategies employed in technological risk governance, MacIntyre's allegory has an unintended double meaning. While over the last quarter century science and technology have proved to be sites of ethical and normative dynamism – provoking new moral questions and problematising received understanding of nature of 'life itself' (Rose, 2006) – a series of contemporaneous technological risk controversies has problematised the role of technical expertise in adjudicating matters of public concern. There is a parallel between MacIntyre's 'disquieting suggestion' and Beck's (1992) conceptualisation of a 'world risk society' characterised by a reflexive critique of scientific expertise and the proliferation of alternative forms of both technical and moral authority. In a society where the production of risk is recognised as structurally linked to modernisation, Beck argues that 'people themselves become small, private alternative experts in the risks of modernisation' (p. 61). Beck's argument – which has been well rehearsed and debated – is not simply that modern societies are subject to a new class of technological risks. Rather he argues that the structural coupling between the production of risks and processes of modernisation – where new threats are literally the 'piggy back products' of technological modernity – provides the conditions for a reflexive critique of modernity. The notion that scientific expertise serves as an unparalleled source of empirical truth and arbiter of cultural meaning is now no longer unchallenged (Doubleday and Wynne, in press). Indeed for Giddens (1994) this form of reflexive modernisation, twined with the overlapping processes of globalisation and de-traditionalisation has ushered in a new era of 'life politics', where questions of individual and collective virtue are paramount. 'Life politics' Giddens argues is characterised by 'disputes and struggles . . . about how we should live in a world where everything that used to be natural (or traditional) now has in some sense to be chosen or decided upon' (p. 90). In this sense MacIntrye's pessimism concerning the revival of questions of ethics and Beck and Gidden's notions of reflexivity revolve around a similar conceptual dynamic. Both concern the relationship between questions of virtue and the nature of moral expertise. In their own way MacIntrye, Beck and Giddens each argue that in the absence of an underlying and shared moral *telos* moral authority becomes reflexively distributed, lacking a basis for collective deliberation of share concerns.

In a curious fashion, this argument helps to articulate a conceptual problem for contemporary critical risk research: how do the 'fragments of morality' deployed as notions of corporate virtue and civic responsibility function as a site of veridiction for the novel materialities of contemporary technological innovation? In the absence of a shared *telos* how does formal risk assessment function as a site for articulating collective anxieties concerning technological change? In the case I have presented in this chapter a set of cultural fault-lines are evident in the responsibilisation of contemporary risk governance. Principle among these is the way in which formal risk assessment functions as an arbiter of relations between the human and non-human world, where the development of preemptive risk management strategies is indicative of the capacity for technological innovation to challenge the constitutional ontology of formal risk assessment. Technologies such as nanotechnology and synthetic biology – where materials and biological systems are endowed with an informally agency – test to breaking point the formal distinction between the 'truth of things' and the 'will of humans' (Latour, 2004). The cultivation of a set of anticipatory sensibilities therefore represents an attempt to render these hybrid materials governable.

The second fault line evident in contemporary risk research concerns the social and cultural legitimacy of formal risk governance. The endowment of scientific and regulatory authorities with a capacity for moral virtue, embodied in regulatory forms that enable the public performance of good governance and corporate self-regulation – are indicative of the capacity for formal risk assessment to resuscitate its political credibility. In place of MacIntrye's notion that notions of ethics and virtue have become a 'mask for almost any face' – initiating a form of morality without *telos* – I suggest that notions of the responsible development of nanomaterials function to make such materials culturally tractable precisely by providing an underlying *telos* for these hybrid technological innovations redolent with the potential for future risks. Such strategies are thus drawn into the informational enrichment of the very materials they seek to govern, producing molecules that 'rich in information about their (global) legal and economic, as well as their chemical relations to other molecules' (Barry, 2005: 64).

Acknowledgements

This chapter is based upon research conducted in two research projects; the European Commission project: *DEEPEN: Deepening Ethical Engagement and Participation in Emerging Nanotechnologies* and the ESRC funded project *Strategic Science: Research Intermediaries and the Governance of Innovation* (RES-061-25-0208).

References

Anderson, B. (2010) Preemption, precaution, preparedness: Anticipatory action and future geographies. *Progress in Human Geography*, **34**(6), 777–798.

Barben, D., Fisher, E., Selin, C., and Guston, D. (2008) Anticapatory governance of nanotechnology: foresight, engagement and integration, in *The Handbook of Science and Technology Studies – Third Edition*. (eds E. J. Hackett, O. Amsterdamska, M. Lynch and J. Wajcman), MIT Press, Cambridge M.A., pp. 979–1000.

Barnett, C., Cloke, P., Clarke, N., and Malpass, A. (2010) *Globalizing Responsibility: The Political Rationalities of Ethical Consumption*, Wiley-Blackwell, London.

Barry, A. (2005) Pharmaceutical matters: the invention of informed materials, inventive life: approaches towards a new vitalism. *Theory, Culture & Society*, **22**(1), 51–69.

BASF. (no date) *Code of Conduct Nanotechnology*. http://corporate.basf.com/en/sustainability/dialog/politik/nanotechnologie/verhaltenskodex.htm?id=VuGbDBwx*bcp.ce. (accessed 4 March 2008).

Beck, U. (1992) *Risk Society: Towards a New Modernity*, Sage, London.

Brickman, R., Jasanoff, S., and Ilgen, T. (1985) *Controlling Chemicals: The Politics of Regulation in Europe and the United States*, Cornell University Press, Ithaca, N.Y.

Brown, N., and Michael, M. (2002) From authority to authenticity: the changing governance of biotechnology. *Health, Risk and Society*, **4**(3), 259–272.

Burchell, G. (1993) Liberal government and techniques of the self. *Economy and Society*, **22**(3), 267–282.

de Goede, M., and Randalls, S. (2009) Precaution, preemption: arts and technologies of the actionable future. *Environment and Planning D: Society and Space*, **27**, 859–878.

Defra. (2007) *Characterising the Risks Posed by Engineered Nanoparticles*, DEFRA, London.

——— (2008) *UK Voluntary Reporting Scheme for Engineered Nanoscale Materials*, DEFRA, London.

Derrida, J. (1995) *The Gift of Death*, University of Chicago Press, Chicago.

Diprose, R. (2006) Derrida and the extraordinary responsibility of inheriting the future-to-come. *Social Semiotics*, **16**(3), 435–447.

Diprose, R., Stephenson, N., Mills, C., Race, K., and Hawkins, G. (2008) Governing the future: the paradigm of prudence in political technologies of risk management. *Security Dialogue*, **39**(2–3), 267–288.

Doubleday, R., and Wynne, B. (in press) Despotism and democracy in the United Kingdom: experiments in reframing relations between the state, science and citizens, in *Reframing Rights: The Constitutional Implications of Technological Change* (ed S. Jasanoff), MIT Press: Cambridge, M.A.

Douglas, M., and Wildavsky, A. (1982) *Risk and Culture: An Essay on the Selection of Technical and Environmental Dangers*, University of California Press, Berkeley.

DuPont, and Environmental Defense (2007) *Nano Risk Framework*, Environmental Defense – Dupont Nano Partnership, Washington.

Dupuy, J. P. (2000) *The Mechanisation of the Mind: On the Origins of Cognitive Science*, Princeton University Press, Princeton.

Ericson, V. E., and Doyle, A., eds. (2003) *Risk and Morality*, University of Toronto Press, Toronto.

European Commission (2009) *Commission Recommendation on a Code of Conduct For Responsible Nanosciences and Nanotechnologies Research & Council Conclusions on Responsible Nanosciences and Nanotechnologies Research*, European Commission, Brussels.

Ewald, F. (1999) The return of the crafty genius: an outline of a philosophy of precaution. *Connecticut Insurance Law Journal*, **6**, 47–79.

Fiorino, D. J. (2010) *Voluntary Initiatives, Regulation and Nanotechnology Oversight: Charting a Path*, Woodrow Wilson International Center for Scholars, Washington, D.C.

Foucault, M. (1990) *The history of sexuality. Vol. 3, The care of the self*, Penguin, Harmondsworth.

——— (1994) On the genealogy of ethics: an overview of work in progress, in *Ethics, Subjectivity and Truth: Essential Works of Foucault* (ed P. Rabinow), Volume 1, The New Press, New York, pp. 253–280.

——— (2008) *The Birth of Biopolitics: Lectures at the Collège de France, 1978–79*, Palgrave Macmillan, Basingstoke.

Franklin, S. (2001) Culturing biology: cell lines for the second millennium. *Health*, **5**(3), 335–354.

Giddens, A. (1994) *Beyond Left and Right: The Future of Radical Politics*, Polity, Cambridge.

Gottweis, H. (1998) *Governing Molecules: The Discursive Politics of Genetic Engineering in Europe and the United States*, MIT Press, Cambridge, MA.

——— (2008) Participation and the new governance of life. *BioSocieties*, **3**, 265–286.

Guston, D., and Sarewitz, D. (2002) Real-time technology assessment. *Technology in Society*, **24**, 93–109.

Jasanoff, S. (1999) The songlines of risk. *Environmental Values*, **8**(2), 135–152.

——— (2005) *Designs on Nature: Science and Democracy in Europe and the United States*, Princeton University Press, Princeton.

Kearnes, M. B., Grove-White, R., P. M., Wilsdon, J., and B. Wynne (2006) From bio to nano: learning lessons from the agriculture biotechnology controversy in the UK. *Science as Culture*, **15**(4), 291–307.

Kearnes, M. B., and Rip, A. (2009) The emerging governance landscape of nanotechnology, in *Jenseits von Regulierung: Zum politischen Umgang mit der Nanotechnologie* (eds S. Gammel; A. Lösch and A. Nordmann), Akademische Verlagsgesellschaft, Berlin.

Keller, E. F. (1995) *Refiguring Life: Metaphors of Twentieth-Century Biology*, Columbia University Press, New York.

Lakoff, A., and Collier, S. (2004) Ethics and the anthropology of modern reason. *Anthropology Today*, **4**(4), 419–434.

Latour, B. (2004) *Politics of Nature: How to Bring the Sciences into Democracy*, Harvard University Press, Cambridge, M.A.

Levidow, L. (2001) Precautionary uncertainty: regulating GM crops in Europe. *Social Studies of Science*, **31**(6), 842–874.

Massumi, B. (2007) Potential politics and the primacy of preemption. *Theory and Event*, **10**(2), DOI: 10.1353/tae.2007.0066.

O'Malley, P. (2000) Uncertain subjects: risks, liberalism and contract. *Economy and Society*, **29**(4), 460–484.

——— (2004) *Risk, Uncertainty and Government*, Glasshouse, London.

Owen, R., and Goldberg, N. (2010) Responsible innovation: a pilot study with the UK Engineering and Physical Sciences Research Council. *Risk Analysis*, in press (30), 11.

Phillips, L., Bridgeman, J., and Ferguson-Smith, M. (2000) *The BSE Inquiry*, The Stationery Office, London.

Rose, N. (2006) *The Politics of Life Itself*, Princeton University Press, Princeton.

Royal Commission on Environmental Pollution (2008) *Novel Materials in the Environment: The Case of Nanotechnology*, HMSO, London.

Royal Society and Royal Academy of Engineering (2004) *Nanoscience and Nanotechnologies: Opportunities and Uncertainties*, Royal Society and Royal Academy of Engineering, London.

Sarewitz, D., Pielke, R. A., and Byerly, R. (2000) *Prediction: Science, Decision Making and the Future of Nature*, Island Press, Washington, D.C.

Shamir, R. (2004) The de-radicalization of corporate social responsibility. *Critical Sociology*, **30**(3), 669–689.

——— (2008) The age of responsibilization: on market-embedded morality. **37**, 1, 1–19.

——— (2010) Capitalism, governance and authority: the case of corporate social responsibility. *Annual Review of Law and Social Science*, **6**, 531–553.

Shore, C., and Wright, S. (1999) Audit culture and anthropology: neo-liberalism in British higher education. *Journal of the Royal Anthropological Institute*, **5**, 557–575.

Strathern, M. (2005) Robust knowledge and fragile futures, in *Global Assemblages: Technology, Politics and Ethics as Anthropological Problems* (eds A. Ong and S. Collier), Blackwell, Oxford, pp. 464–481.

Sutcliffe, H., and Hidgson, S. (2006) *Briefing Paper: An Uncertain Business: The Technical, Social and Commercial Challenges Pesented by Nanotechnology*, Acona, London.

United Nation (1992) *Report of the United Nations Conference on Environment and Development*, United Nations, A/CONF.151/26 (Vol. I), Rio de Janeiro.

Vogel, D. (2005) *The Market for Vırtue: The Potential and Limits of Corporate Social Responsibility*, The Brookings Institute, Washington, D.C.

von Schomberg, R. (2010) Introduction: understanding public debate on nanotechnologies, options for framing public policy, in *Understanding Public Debate on Nanotechnologies Options for Framing Public Policy* (eds R. von Schomberg and S. Davies), Directorate-General for Research, Science, Economy and Society, European Commission, EUR 24169 EN, Brussels.

Wilsdon, J., and Willis, R. (2004) *See-Through Science: Why Public Engagement Needs to Move Upstream*, Demos, London.

Wynne, B. (1982) *Rationality and Ritual: The Windscale Inquiry and Nuclear Decisions in Britain*, British Society for the History of Science, Chalfont St Giles.

——— (2006) Public engagement as a means of restoring public trust in science – hitting the notes, but missing the music? *Community Genetics*, **9**(3), 211–220.

PART 3
Ethics in Risk Research

CHAPTER 8

Ethical Risk Management, but Without Risk Communication?

Stuart N. Lane

Institut de géographie, Faculté des geosciences et de l'environnement, Université de Lausanne, Lausanne, Switzerland

Introduction

Adams (1995) contrasts approaches to risk management along two axes (Figure 8.1). The first relates to assumptions regarding the nature of the risk as lived, ranging from risk that is defined by experience through to risk that is experienced by definition. The second relates to the nature of the desired response to those risks, ranging from the assumption that individuals can and should manage the risks that they are exposed to themselves, through to the notion that it is only in the collective or aggregate that responses to risk can be achieved. In some senses, it is possible to write a history of risk management in the more developed world which is about a progressive drift across the space defined by Figure 8.1, towards what Adams (1995) calls the hierarchical model of risk. In response to a combination of the experience of catastrophic events and the growing faith that science and technology, progress, can be a means of preventing such events, we have progressively created a perception that risk is something that can be managed; and that this management has to happen centrally. We require centralised institutions, ones that are capable of reviewing a situation, invoking the technologies required to make knowledge such as observations or predictions (see Klauser and Ruegg, this volume): of the existence of a risk; the presence of an impending event; or the effects of interventions that reduce risk by forcing events to happen less frequently. The spatial scale of such endeavours can be significant and in many cases can only be realised through collective endeavours. In my field of research, flood risk, large scale infrastructure projects, national scale warning systems and the provision of insurance and reinsurance cover are beyond smaller scales of provision. To argue that such endeavours could be anything other than large in scale is not my intention in this chapter.

Critical Risk Research: Practices, Politics and Ethics, First Edition.
Edited by Matthew Kearnes, Francisco Klauser and Stuart Lane.

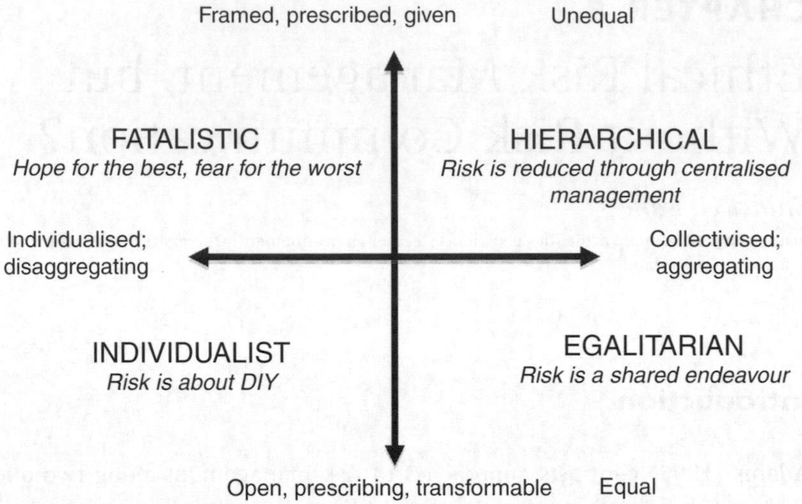

Figure 8.1 Approaches to risk management (modified from Adams, 1995).

Rather, I want to think through one of the implicit assumptions made in the hierarchical approach: that science exists to assess risk; that statutory bodies, what Callon (1999) calls intermediaries, exist to manage those risks; and, in turn, that those who live with risk are expected to conform to the system of risk management to which they are exposed. First, I argue that this hierarchical model raises difficult ethical questions; ones that are rarely considered because of an in-built assumption that centralised risk management exists in the collective good, to identify, to manage and then to target communication to those who are at risk, so that their exposure to risk is reduced. Second, I will argue that these ethical questions are bound to asymmetries in the access and credence given to different sources of knowledge. It is this asymmetry that leads me to consider how, in the final sections of the chapter, risk management can be redemocratised not through better communication, but co-production. In this chapter, I use the general example of flood risk in the United Kingdom (UK) and the specific case of Ryedale, in North Yorkshire, a site where I have recently been involved in a project to co-produce knowledge.

Risk thermostats, risk hierarchies and democratic accountability

Figure 8.2 shows a series of conceptual models that Adams (1995) labels the 'risk thermostat'. Adams (1995) argues that as individuals we balance an in-built propensity to take risks against perceived dangers (Figure 8.2a).

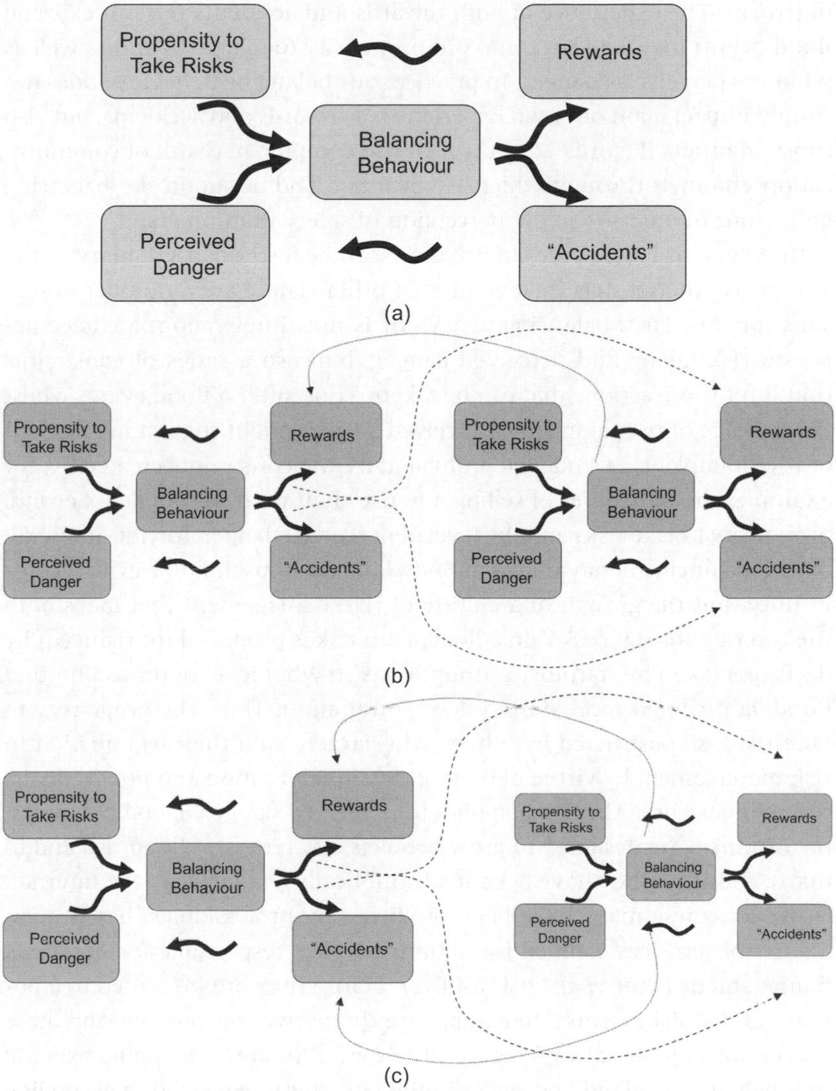

Figure 8.2 The 'risk thermostat' after Adams (1995): 2a – the basic thermostat; 2b – interactions between thermostats; and 2c – asymmetry in interacting thermostats.

In turn, taking risks can result in both rewards (the beneficial effects of an action, but where no risk is materialised) and accidents, where the negative effects of taking a risk are experienced. The decision to live in a floodplain, for instance, may be taken because the perceived danger is low because we are aware of the presence of flood defences (White, 1942), even though we may have a low propensity to take risks. We may perceive high rewards from living in a floodplain such as ease of access to infrastructure, which also tends to be found in floodplains (e.g. the major road network). The experience of both rewards and accidents (e.g. an extreme flood event) may feed back into our propensity to take risks and as well as what we perceive as danger. In practice, our balancing behaviour does not simply impact upon our own experience of rewards and accidents, but also those of others (Figure 8.2b), through the complex network of communication channels through which we evaluate and act upon the balancing behaviour of others (e.g. the perception of safety in numbers).

However, as Figure 8.2c shows, there can be a strong asymmetry in the size of risk thermostats relative to each other. There are a number of reasons for this. First, balancing behaviour is not simply about a choice between risk taking and perceived danger, but also a series of constraints that limit those actions that can be taken. Thus, after a flood event, whilst the balance of risk taking and perceived danger might suggest moving out of the floodplain, this may be prohibited by the constraints created by, for example, the difficulties of selling a house that has been flooded. Second, propensity to take risks and the level of perceived danger are not simply set by the balancing behaviour of an individual, a household or even a community, but the growth of a culture of risk management that maps onto the top right of Figure 8.1. In a floodplain, risk is produced (or reduced) by decisions taken by statutory authorities as to what level of protection that floodplain should receive from flood inundation. Thus, the propensity to take risks is constructed by others, who largely gain their legitimation in risk management by virtue of their professional position and not their personal experience. The same applies to perceived danger. Flood risk maps, for instance, are designed to grow people's awareness of flood risk and in many countries they have been made publically available via the Internet. However, these maps have been produced by professionals, based upon a suite of practices codified by statutory bodies responsible for flood risk management (Lane *et al.*, 2011a). Even before they are presented to a potential floodplain victim, these maps are themselves perceptions and these perceptions are not always correct. The entire flood risk mapping exercise that began in England and Wales in the late 1990s perceived that flooding was caused by water courses designated by legislation as main rivers, and not direct runoff from hillslopes or rivers designated as 'ordinary water courses' (generally smaller rivers, but with no explicit size-related definition). Extensive experience of flooding by direct runoff and non-main rivers in the summer of 2007 unsettled this assumption. It also revealed that the perceptions that we develop as to whether or not we are at risk, our own balancing behaviours, have the potential to be shaped by the perceptions of others.

Here, I want to argue that a critical point arises from asymmetry in the risk thermostat. The assignment of risk management responsibilities to institutions is clearly necessary in many, if not all, cases. The scale of some undertakings (e.g. infrastructure), and the technologies that underpin them (e.g. Klauser and Ruegg, this volume), necessitate institutions with significant resource (financial, human etc.). However, such assignment has to be viewed critically if it leads to asymmetry in the relationship between risk thermostats and that asymmetry matters. The balancing behaviour of those who manage risk through their professional responsibilities will be different to those who live with risk, and I develop this point below. But those who manage risk are also able to express the outcomes of their balancing much more explicitly because they have statutory authority over the decisions taken over precisely how risk should be managed. In theory, risk management decisions should be based upon an objective analysis of the associated risk, such as a cost-benefit analysis; but evidence shows that in practice risk analysis can only proceed when it has been framed by a series of policy decisions as to how that analysis is undertaken (Lane *et al.*, 2011a). Thus, in effect, we have vested and, crucially, entrusted the balancing behaviour that defines how we experience risk to those institutions that manage risk. Again, in theory, such institutions remain accountable to us through the political system through which they are regulated. Take the example of the Environment Agency of England and Wales (the E.A.). The E.A. is a non-departmental public body responsible to the Department of Food and Rural Affairs (DEFRA) and hence to the UK government and democratically-elected Members of the UK's Parliament. However, the EA is a large and complex organisation, undertaking many duties. Difficult questions follow over the extent to which its notional democratic accountability is indeed realised in practice, as many of the actions it takes may not be immediately open to scrutiny.

Knowledge and living with flood risk

In a case study of flood risk in Pakistan, Mustafa (2005) describes how accounts of hydrological hazards as science-centred technical challenges have come to dominate at the expense of other accounts, such as those that are experiential, and grounded in the understandings of those who live with risk. Mustafa argues that for a floodplain dweller, the 'hazardscape' comprises a balancing act regarding the multitude of risks that are experienced as part of the business of day-to-day life: hazards are ever-present, hydrological hazards are only ever intermittently present. Such experiences are bounded both historically and spatially and Mustafa used the notion of the hazardscape as something that:

> Fuses the material and discursive aspects of how hazardous spaces are produced, contested, and struggled over (Mustafa, 2005: 570).

This is in marked contrast to situations where institutions or statutory authorities are responsible for risk management. Here, the balancing act is much less likely to be some kind of geographically and temporally contingent trade off involving the multitude of risks thought to exist. Rather, it will involve a balancing act that, whilst still bounded geographically and temporally over much larger scales, involves a comparative prioritisation of different geographical locations for a single risk. For instance, an institution's balancing activities may be concerned with the optimal spatial allocation of scarce resources as compared to the potential costs of not investing. The result is two very different kinds of position with respect to risk, one concerned with balancing the range of risks being faced in a given location and the second with balancing the range of geographical locations in relation to a particular risk. Of course, there are hidden links between them: if the spatial prioritisation of investment in risk reduction is not right then this may feed back into the set of risks being faced by an individual if a major flood event has national economic consequences. But there are also explicit links between these two kinds of balancing act because institutional activities are expressed both materially (e.g. flood defences) but also discursively such as through consultation and risk communication or when knowledge about flood risk impacts upon day-to-day lives.

In giving authority to statutory bodies to manage risk on our behalf, we inevitably transfer to them responsibilities for the generation and management of knowledge. In practice, knowledge of flood risk may be developed by scientists within institutions; or undertaken by scientists working in consulting organisations employed by institutions (Landström *et al.*, 2011). In the latter, institutions becoming intermediaries (Callon, 1999) or contact points between the knowledge generated by specialists and those who are required to live with the consequences of that knowledge. Institutions need this knowledge as a means of strategically identifying where flood risk is likely to be found, developing detailed assessment of risk where it is thought to be high, evaluating the costs and benefits of interventions to reduce those risks and delivering and maintaining those interventions. In addition, institutions may gain responsibilities for forecasting the manifestation of risk as individual events. Both general and event-specific risk management lead to a strong emphasis on communication between intermediaries as the holders of knowledge and the public, often under the classic assumption that the public are 'deficient' in knowledge and understanding whilst scientists, and the intermediaries that they inform are 'sufficient' (Sturgis and Allumn, 2004). This kind of communication is discrete in the sense that it tends to focus on the products of knowledge, when they have become stabilised after scientists and intermediaries have agreed them.

This approach to knowledge generation, management and communication can be illustrated by the EA's activities in England and Wales. The EA is responsible for the strategic identification of flood risk associated with main rivers (see above) as well as coastal inundation. To do this, it maintains a network of rainfall and river flow gauges. It commissions scientific consultants under framework agreements (see Landström *et al.*, 2011) to analyse the data from these gauges and to apply them in mathematical models for flow routing and floodplain inundation for strategic analysis, detailed risk assessment, the design of interventions and development of flood forecasting models. The EA is then the intermediary between those who generate knowledge and those who communicate it to those at risk. This communication strategy involves a number of methods. It maintains various targeted communication systems (e.g. automated telephone calls) for rapid dissemination of warnings during events. However, it also embarks upon wider communication strategies, such as through the publication of strategic flood assessments on-line and supported by a postcode finding system, awareness raising events in flood risk areas and consultations over the interventions it plans to take to reduce flood risk.

Ultimately, what underpins this knowledge mirrors the position that the EA is required to take as a statutory authority. The knowledge it generates is comparative and geographically-extensive, based upon strategic assessments over large areas which in turn focus in on more localised assessments, although, even these may cover a whole town or large parts of a city. The kind of knowledge that is being used here is surveillant in the sense that it is created using mathematical models of flood inundation coupled to boundary conditions (e.g. river discharge and river bathymetry) and, crucially, remotely-acquired floodplain elevation data. Such analyses may be further supported by property databases. Whilst the consultants who perform the modelling activity may visit the sites they are modelling (Landström *et al.*, 2011), this is rarely to look at the extent to which individual properties have been correctly represented.

This knowledge, and especially given its dissemination on-line, meets those who live with flood risk in ways that can be controversial and there are a number of reasons for this. The controversy is not simply about matters of policy and decision-making but also knowledge. The models that the EA commission are generally correct but wrong in terms of their detail (e.g. Bradbrook *et al.*, 2004): such models are commonly assessed by considering the level of agreement (number of cells predicted correctly as inundated plus the number of cells predicted correctly as not inundated) or the level of disagreement (number of cells predicted as inundated when dry plus the number of cells predicted as dry when inundated), as compared with the total number of cells considered. Aside from technical issues regarding precisely how these indices are calculated, the percentages are never 100% (for agreement) or 0% (for disagreement) and whilst the models are generally right their detailed predictions are wrong (e.g. Yu and Lane, 2006a, 2006b). Yet it is precisely this detail that a floodplain dweller will experience. To this problem can be added a number of other artefacts such as the need to provide consistency between different geographical places. As a result, the flood maps commonly show a particular return period, sometimes called a 'design event', which may bear no resemblance to the actual size of the events that have actually been experienced by flood dwellers. The maps are often produced with a purpose, which can also impact upon what they show. The initial flood risk maps produced for England and Wales excluded flood defences, labelled the 'undefended floodplain', such that large areas of floodplain were predicted as being at flood risk even though they were behind flood defences. Such maps were an approximation of the true risk from flood inundation (i.e. because any flood defence carries a probability of failure) but often showed very little resemblance to the knowledge of those who lived behind those flood defences. Regardless of which risk should be shown on the maps (the true or undefended floodplain risk or the risk assuming defences hold), the decision to use the undefended floodplain risk reflects a particular logic for the flood maps: the institutional need to grow awareness of the risk of flooding in the hope that a range of individuals would respond (planning decisions, flood proofing of properties). Those labelled at risk had no choice in the publication of these maps even though the maps are now used routinely, by solicitors, to undertake searches in relation to property sales. Similar technologies have been used by insurance companies to improve the pricing of insurance policies. The knowledge held institutionally about those properties at risk from flood inundation is surveillant in the sense that it is generated without any permission from a business or householder, despite the potential implications of that knowledge.

Three observations follow. The first is an ethical one. On the one hand, and as noted above, some flood risk can only be managed through institutions capable of undertaking large-scale risk assessment and able to draw upon necessary technical methodologies, notably forecasting tools. On the other, those to whom this knowledge pertains have little or no control over how it is used, even whether or not it should be used at all, even though it may have profound implications for their well-being. The ethical asymmetry surrounding who takes balancing behaviour is compounded by an asymmetry in the knowledge that is allowed to be admissible in flood risk management. As this knowledge becomes entrenched in institutions, then a culture of risk dependency develops. Those for whom risk management is a professional concern, generate knowledge about risk. This risk is used to reach management decisions, something that progressively removes risk from our day-to-day lives, sanitising the landscapes within which we live. We lose the experience and knowledge we need ourselves to learn to how to live with risk, not just the experience of risk itself, but also the experience of coming to know about risk. In turn, this makes us more dependent upon those in whom we have invested responsibilities for risk management and knowledge provision, knowledge that we no longer are able to obtain through our own experience. The asymmetry becomes encoded into our day-to-day lives as we become ever more dependent upon institutions to generate knowledge for us.

The second observation compounds this ethical observation in that the ethical asymmetry is not simply about who has the capacity to acquire knowledge but who is deemed as entitled to access knowledge. Consider, for example, recent policies adopted by the EA with regard to the licensing of data. Because of the privileged position we tend to give data as a form of knowledge in decision-making (i.e. best available evidence), data are extremely powerful. The majority of the funding that the EA receives is from the UK Treasury and hence from those who pay tax to the UK government. Some of this funding is used to support the network of river gauges and hence the data used in flood risk management. However, the EA now licenses these data for non-commercial use with the specific clause (3.5):

> Any intended use of Information must not represent a risk of: being misleading to anyone you are allowed to pass the Information to; detriment to the Agency's ability to achieve its objectives; or detriment to the environment, including the risk of reduced future enhancement; or being prejudicial to the effective management of information held by the Agency; or damage to the Agency's reputation.

This clause illustrates, at once, the power that data are thought to have (i.e. through the need to regulate its use) and the prohibition of the use of this power in a way that might undermine the organisation holding those data themselves. Through the regulation of data in this way, the EA is able to maintain its position in the asymmetric relationship between its role as a risk manager and those who are subject to the effects of the decisions that it takes. For the EA, there is no choice in evolving to this situation. On the one hand, it has to make difficult decisions about what it is and is not allowed to fund in relation to flood defence. On the other, it has to manage its position with respect to those who we have elected, to keep itself politically-accountable. Damage to reputation is a very real issue. Nonetheless, there remains a fundamentally ethical concern that a body is able potentially to regulate access to the power that comes from knowledge, restricting the use of the knowledge by those who have funded the acquisition of that knowledge in the first place, and whose livelihoods may be profoundly impacted upon by the decisions taken using that knowledge. It emphasises a critical point: in a culture that places a central emphasis upon evidence-based decision-making, knowledge becomes a critical commodity. The ethical concern is not simply over the asymmetry in access to knowledge but also over the regulation of its use.

The third observation relates to differences in the framing of knowledge that is deemed important to institutions. The use of return periods, the inclusion or otherwise of flood defences and models that do not get the detail of inundation extent right contain at once both a logic and a tension. The knowledge held is that which is required to allow the institutions to function: knowledge that allows the comparative evaluation of locations; and the subsequent prioritisation of interventions. The tension arises during events, when that knowledge is brought into sharp scrutiny. Following Rozario (2007), events allow those who experience them, directly and personally, to challenge institutional calculations, forcing institutions to construct those events as exceptions. As Callon (1999, 85) notes: most publics possess *'specific, particular and concrete knowledge and competencies, the fruit of their experience and observations'* (Callon, 1999, 85). The literature regarding the relationship between science and technology emphasises the supposed differences in the form of knowledge held by scientists and publics (e.g. Collins and Evans, 2002). However, this literature fails to recognise that all of us learn through experience, no less so in relation to risk management where the means by which we assess risk are directly a product of those risks that we experience (e.g. Slovic, 2000). The kind of risk experienced by a professional manager of risk is likely to be the statutes and institutional norms within which they work; the computer models and analytical devices that underpin the science that they do or which they commission; the maps and other artefacts that they produce; and possibly, but by no means certainly, their exposure to the communities that have to live with risk. That of the floodplain dweller will be the highly infrequent occasion when water enters their property or impacts upon their ability to follow their normal day-to-day activities, coupled to what is now critical, but often overlooked, in relation to the long-term emotional effects of experiencing flooding (Ohl and Tapsell, 2000). In other words, in addition to approaching risk from different positions, the risk manager also approaches risk with very different kinds of experiential knowledge to that of a floodplain dweller.

Such knowledge may be classified comparatively; the geographically extensive and comparative kind of knowledge that a risk manager holds; and the geographically and historically specific knowledge of a floodplain dweller. Traditionally, we assume that such a position provides the risk manager with a suitably privileged position with respect to risk assessment. The risk manager can remain independent from the supposed biases and emotions that make risk a 'matter of concern' (Latour, 2004) for particular individuals in particular places. The flood risk manager's decisions can be based upon sound, science-based policy. However, flood risk science can only work if it is constrained by a series of policy decisions (Lane *et al.*, 2011a), a critique of naïve scientism. Whilst these policy decisions may appear to allow for the routinised comparison of knowledge in a way that implies objectivity, it is equally clear that they can lead to questions regarding social justice (Johnson *et al.*, 2007) and that they are not enough to constrain the scientific practice of flood risk modelling (Lane *et al.*, 2011a). When opened up to scrutiny during events, the divergence between the kind of knowledge generated and held by institutions and that of floodplain dwellers becomes the basis of controversy. A simple question arises: can we find ways of managing flood risk that are simultaneously more ethically acceptable in relation to balancing behaviour and knowledge but which also can make use of the energy contained within such controversies? The challenge is to rethink the position of publics in risk management.

Redemocratising risk management

Callon (1999) contrasts two broad responses to his observation that most publics themselves possess knowledge of the problems that they are subject to: consult and debate with them over that knowledge; and, more radically, work with them to co-produce new knowledge. They differ from the deficit model in that they progressively reposition publics with respect to knowledge. Under consult and debate, it is recognised that the public have both the capacity and the entitlement to deliberate knowledge, even if such knowledge, itself, is generated by others. Under co-production, the public is not only assumed to have considerable capacity to generate meaningful knowledge; it is also recognised that to produce knowledge without the active involvement of those who will be subject to that knowledge is likely to produce either a partial or even an incorrect understanding.

The consultation and debate model introduces the notion that, for those who have a stake in a problem, and where that problem is being approached through the generation of knowledge, then knowledge should be taken as provisional until there has been both consultation and debate over it. The point is not just consultation (i.e. 'what do you think?') but also deliberation (i.e. 'is the knowledge right?') and there is now a long list of approaches advocated for doing this (e.g. the 'extended peer review' of Funtowicz and Ravetz, 1991, 1993; and Yearley, 2006). This consultation and deliberation may be extended upstream to address the way that problems have been framed in the first place (e.g. Wynne, 2003; Costanza and Ruth, 1998; Cockerill *et al.*, 2006). There is no doubt that, where adopted, consultation and deliberation opens up decision-making to public scrutiny. However, there is now a sufficient body of evidence for the theoretical appeal of the approach to be contrasted with what happens when it is pursued in practice.

In practice, consultation and deliberation remains detached from the process of knowledge generation itself. Responsibility for reformulating knowledge generated rests with those who generated it in the first place. First, there is evidence that at least some consultation and deliberation type engagements are little more than exercises in legitimation (Harrison and Mort, 1998; Stirling, 2008), criteria-compliance (Stirling, 2008), processes to be worked through to tick boxes rather than impacting on outcomes. Deference to those who generate knowledge is retained and perhaps even reinforced because of the apparent legitimation that consultation brings to it. Consultation and debate reinforces the ethical dilemma raised by the hierarchicalist approach to risk management more generally because it becomes a means of sustaining the position of institutions in the hierarchy whether they have generated knowledge themselves or commissioned others to generate it. Second, debate and consultation represents an internal contradiction (Lane *et al.* 2011b). On the one hand, it recognises that knowledge may be provisional and uncertain and therefore debateable. It unsettles the role that we might want knowledge to play in decision-making as that knowledge is brought under closer scrutiny and we discover that we should not necessarily and automatically trust the knowledge generated by others (Collins and Evans, 2002). On the other hand, it reveals nothing of the practices by which that knowledge is produced, and what leads those who produce that knowledge to trust it. Trust in knowledge is generated through activities that are commonly experiential (e.g. learning through experiment), judgemental (e.g. peer review) and, above all, practiced (Lane, 2011). If it is the practice of generating knowledge that develops the trust that a scientist or consultant holds in their own work, then the only means by which others might share that same sense of trust may be through being involved in the same practices and, crucially, through coming to understand their limits and possibilities, perhaps ultimately changing those practices. Third, Callon (1999) argues that consultation and deliberation runs the risk of becoming singular in direction in which the only knowledge subject to testing is that generated by those given responsibility for generating it. For Callon, all knowledge, including lay or local knowledge, should be put to the test. Whilst this implies a repositioning of those who generate knowledge in the decision-making process it is, at least in theory, not a particularly exceptional development: the essence of scientific enquiry, for instance, is putting ideas to the test. It is exceptional, even radical, because it implies that those who should be entitled, even required, to do the testing are not simply those who are given that privilege through their position in the institutional hierarchy, such as scientists or employees of institutions. Justice, then, is not simply a summative outcome, but one that is achieved through a just set of procedures.

For Callon (1999), co-producing knowledge is the obvious response and he envisages the co-production of knowledge as a process of dynamic collective learning, involving those for whom an issue is of particular concern, whether as a result of their professional position (their 'certification' following Collins and Evans, 2002), their personal position with respect to a local issue or their experience of an issue. Institutions and the knowledge that they generate, or which they ask others to generate, are still critical. The difference is that they are no longer given special privilege over knowledge produced or held by others. The approach here is not a 'hoover', by analogy with a vacuum cleaner; i.e. it is not simply about assimilating the knowledge of publics in ways that supposedly make institutional knowledge somehow better. Not only does such a view raise the same kinds of ethical considerations as consultation and debate approaches, but it overlooks evidence that, for at least some risks, publics can have a highly developed understanding of the same general concepts that scientists hold (Lane *et al.*, 2011b). Rather, expertise is seen to be less differentiated (e.g. on general versus local grounds) and more distributed than is commonly assumed.

Co-producing flood risk management

The above ideas have some appeal in theory, but it is perhaps less clear as to how they might be put into practice. In this section I outline one experiment concerned with co-producing flood risk management and the critical findings that came from it. In England and Wales, the Department of the Environment, Food and Rural Affairs published a report in 2008 that specifically considered broader questions of procedural justice in the work of the Environment Agency. DEFRA saw procedural justice as providing 'opportunities to participate in deliberations and influence decisions' (2008, 18) and specifically explored the EA's view of public engagement. DEFRA concluded (2008, 34) that whilst the EA's Vision 2000 emphasised procedural justice (openness, partnership, participatory decision-making) 'chiefly it is consultation rather than more deliberative processes that is mentioned.' In relation to the EA's 2006-11 Corporate Strategy, it notes:

> With regard to the public, the EA talks about being an effective communicator, presenting information to the public and listening to their views; rather than involving them in participatory decision-making processes (DEFRA, 2008, 35).

In relation to the EA's flood risk management strategy 2003-4 to 2007-8 again, whilst noting that the EA's strategy emphasises procedural justice, in practice:

> The communication envisaged appears to be consultation and information exchange rather than participatory processes (DEFRA, 2008, 36).

Thus, DEFRA (2008, 173) concluded 'Stakeholder engagement tends to be focused on consultation and the provision of information; rather than participatory stakeholder engagement per se'.

Much of this is reflected in the Environment Agency's implementation of its public engagement responsibilities through a tool kit *Building Trust With Communities* (Environment Agency, 2004a, 2004b), which sits within the Environment Agency's Corporate Affairs section. This is based upon 12 principles (Table 8.1). Principle 3 illustrates the emphasis upon filling knowledge deficits; Principle 4 the notion that trust is a matter of

Table 8.1 *Building Trust with Communities'* 12 core principles (Environment Agency, 2004a, 13–14)

1	Fair for all. Every person who has an interest in, or who could be affected by, the issues under discussion must be encouraged to take part.
2	Be clear at the start about what changes the Environment Agency can or cannot promise and be clear about the mechanisms of the decision-making processes.
3	Ready information. Be sure you give people as much information as possible and explain where information is missing or is uncertain.
4	Show respect for diverse views and cultures by making sure that minority views are taken on board. Respect interested parties and taxpayers by making sure that your work with local communities is seen as a priority and has widespread support from the community. This is your opportunity to build trust by being courteous, empathic and helpful.
5	Feed back. Use existing channels to make sure that you report back to all interested people as fully and as quickly as possible.
6	Take action. Put final decisions into action as soon as possible. This will strengthen participants' belief that their involvement was worthwhile.
7	Each time there will be lessons to be learned for both the Environment Agency and the community groups, building mutual understanding, trust, respect and relationships. Some initiatives will fail but they should be seen as valuable contributions as they provide fresh insights.
8	Stand alone. The Environment Agency needs to remain independent throughout the exercise.
9	Common approach. The Environment Agency needs to convey that it is guided by principles that are based on objective professional standards and must be seen to apply these standards across different contexts.
10	No time wasters. Make effective use of time and funding resources for all.
11	Balancing act. The amount of time spent on a project should depend on how important it is.
12	The bigger picture. The aim of everything the Environment Agency does is to improve the environment.

communication; and Principle 5 the separation between decision-making and communication of that decision-making. The translation of these principles into flood risk science (Environment Agency, 2005, 16) reinforces the deficit model and how filling this deficit is the means of engendering trust:

> Always explain what is being done and why a particular position has been chosen.... Explaining every decision... may seem excessive but if trust is to be built with the communities that the Environment Agency serves, it is essential that people understand what the Environment Agency does and how and why it is done.

The last point emphasises the fundamental limit of what can be achieved by consultation. When I started working on a project, along with other natural scientists and social scientists, in the town of Pickering, North Yorkshire, major flood events had occurred four times in fewer than ten years. There was no flood protection and, in the words of the area EA flood risk manager, the town would be unlikely to make the list for further consideration until 2018. How, in this situation, does an institution consult with a local community over the outcome of a knowledge generation process (i.e. a strategic flood risk assessment and subsequent cost-benefit analysis of design options) that will deliver no flood protection for the town? In our case, in 2007, the EA actively steered us towards Pickering, and provided us with some of the data that we needed, but were willing to step back and see what we achieved precisely because Pickering had become, for them, a problem that was hard to progress.

The methods that we adopted in Pickering have been fully reported (Lane *et al.*, 2011b) and so the aim of this section is to summarise some of the more generic characteristics of our co-production of knowledge. Central to our approach was to recognise that, as science recognises routinely, generating new knowledge about Pickering would require us to generate ideas that we could put to the test through 'Environmental Competency Groups'. There were five distinctive elements to our co-production of knowledge. First, we emphasised knowledge-production as well as the knowledge produced itself. This makes the approach distinctive from approaches to public engagement like focus groups, where the focus is on what people think or believe about the products of knowledge. Second, this knowledge production, as far as was possible, was a distributed process, undertaken for a period of time. Third, we emphasised the role that an event can play in bringing into sharp focus the prevailing framings associated with a problem, and the people and things bound to it, so mobilising and enabling those people and things otherwise excluded from the process. Fourth, we sought to bridge the traditional division between academic focus national level flood research projects and the focus of local people on their particular flooding problems. In doing this, the intermediaries normally responsible for obtaining and communicating flood risk knowledge were not a part of our process. Fifth, we did not set out to be representative of pre-existing stakeholder groups (e.g. land managers, flood victims, public bodies, local politicians and officials etc.). Explicitly, this recognises that there is no such thing as a 'representative' group because any attempt at representing complex social-economic-political composition of communities is itself an act of framing around a preconceived notion of what that composition is.

The meetings of eight local members, five academic members, a facilitator and a recorder took place six times every two months from September 2007 to July 2008. The facilitator ran each meeting but also played an important facilitative role between meetings in making sure that local members were able to maintain an appreciation of the principles outlined above. Although the initial work focused on the meetings, these themselves rapidly led to between meeting events and activities, involving different combinations of academic and local members. The group decided to give itself a name, the 'Ryedale Flood Research Group' (RFRG). These activities included data collection, a reading group to look at Consultancy Reports and so on. All of the six main meetings were attended by all members, except for two occasions when, on each, one member could not attend due to family/personal commitments. Initially, meetings were structured by the Facilitator, but as the collective competence of the RFRG developed so the meetings started to structure and direct themselves. The progress of the work is summarised in Table 8.2.

This experience of co-producing knowledge regarding flood risk management allows a number of observations to be made. First, through this way of working, the mathematical modellers in the project, including myself, were forced to turn away from the established methods that they were using to develop a new model suited to the needs of the specific problem that they were working with (Landström *et al.*, in press). We had to dissociate not just from the technologies that we had access to but also the network of scientific allies, including models, that made our work comfortable. Central to this dissociation was the Group's frustration that the full range of flood reduction solutions had not been explored (Lane *et al.*, 2011b), forcing us to look at our own practices in relation to what these other risk reduction solutions might need.

Second, and perhaps more surprisingly, the Environmental Competency Groups revealed a very rich understanding of hydrological and hydraulic processes distributed across all members of the group. Indeed, the knowledge brought to the meetings by local members was not substantially different to that held by academics in the group except in two subtle ways: the language used to express this knowledge was vernacular

Table 8.2 Summary of RFRG activities.

CG1. Starting the Narrative. Introductions; explanation of the wider project, discussion of 'brought objects' within break out groups (we were all asked to bring an 'object' to illustrate our connection to flooding); discussion of flooding as a matter of concern; co-production of a history of flooding in Ryedale.
CG2. Working with Crossing Points 1. Discussing Environment Agency flood inundation maps in break out groups and plenary; plenary discussion of how these are produced, introducing models; marking up of maps in terms of solutions for reducing flood risk; plenary discussion; discussion of what needs to go into models, hydrology and hydraulics.
CG3. Trying Things Out. Round up of activities from all; plenary discussion of computer modelling; trying out the 'bund model' to look at upstream storage in break out groups; plenary discussion, including limitations; identification of research needs, including data, and plans for collecting it.
CG4. Working with Crossing Points. Round up of activities from all; reports on data gathering; discussion of wider developments in flood risk policy; working with video imagery produced by local member; discussion of river maintenance and its relationship to flood risk; discussion of next phase of modelling; planning for what the RFRG might go on to 'produce'
CG5. Trying Things Out. Round up of activities from all; plenary discussion of hydraulic models; working with a hydraulic model to explore maintenance impacts on water level in break out groups; plenary discussion of findings; reflection on flood risk policy in relation to maintenance; decision-making over what the RFRG would produce; allocation of tasks.
CG6. Where next? Round up of activities from all; plenary discussion of how to intervene in Ryedale flood risk management more generally; finalisation of a 'going public' event; completion of ethical and data permission agreements.

(see Table 4 in Lane *et al.*, 2011b); and it was neither structured nor classified formally. The recordings of meetings showed that the means by which local members had acquired this knowledge was experience, notably observation, and that this was no different to the ways in which a scientist might develop their knowledge through observation. The level of sophistication in this local knowledge provides a *prima facie* reason for not seeing managing risk as being concerned with educating a public somehow deficient in knowledge. But, it also revealed the need to debate, to challenge and to structure local knowledge. It was actually the computer models, both developed from the collective deliberations of the group and which were used during the meetings in parallel, mixed academic-local teams, which performed this role. Indeed, it was only through working with material objects, like the computer model, that all members of the Group gained a sense of their own knowledge. Lane *et al.* (2011b) reported a substantial change in the extent to which local members were willing to put their knowledge to work in criticising other knowledge primarily through the process of building and running the model. They discovered what they knew and could come to know and their growing sense of knowledge eventually resulted in substantial correction to the modelling activities. The growth in self-belief was also accompanied by a growth in self-criticism as the solution that came out of the modelling activity was exposed to interrogation, such as in terms of its downstream implications. In other words, and in the same way that academic members had dissociated themselves from their normal networks, so local members dissociated themselves from their own networks, such as the concern that Pickering should be protected from flooding at whatever cost.

Finally, the group's sense of collective competency grew to the point at which a decision was taken that the otherwise confidential deliberations of the group should be made public through an active intervention in flood risk management in Pickering. An exhibition of our work was mounted in October 2008 and attended by almost 200 people. Over the same period, Ryedale became the focus of a bid to DEFRA to be a demonstration project in using rural land management to reduce flood risk, of which the solutions generated by the RFRG would be further explored. This bid was successful and the project is now underway. At the point at which the RFRG went public the model and the group's findings had not been subject to full independent testing. The intervention may appear to have been premature. However, this overlooks what the RFRG had really achieved. Through the process of deliberation and knowledge generation it had created a new collective competence, one firmly aligned to the group, and one able to make a public intervention that would go on to unsettle existing institutional practices in the case of Pickering. The motivation, then, of the group's knowledge generation activities had come to mirror that of those institutions' given statutory responsibility for knowledge generation: to sustain its position through a grounding in its own collective competence, the knowledge that it had generated.

Ethical reflections

My motivation for writing this chapter was a sense that there has been a deeply unethical trend that has accompanied the institutionalisation of risk management, the drift from bottom left to top right in Figure 8.1, and one that has resulted in dedicated professionals serving the needs of an institution or coalition of institutions, public and/or private (e.g. Klauser and Ruegg, this volume), more than those whom we have come to expect to be provided for by those institutions. That this should have happened is not surprising. Even without the development of social media, as we become more dependent on institutions to manage risk for us, so those institutions become more exposed to popular protest and political intervention in those inevitable instances where risk management appears to have failed. If institutions are given responsibility for managing risk, events may simply become (mis-)perceived as arising from institutional failure. Institutions have had to respond to this through investment in public relations departments, careful monitoring and control over both data and the private providers of data, lobbying activities with government and development of close relationships with democratically elected officials. Serving institutional needs becomes very real and very important for the ongoing existence of those institutions and the personal and social functions that they sustain.

In such a situation, there remains a constant need to reinforce, perhaps to grow, amongst those professionals involved in technical approaches to risk management, including myself, a sense of 'moral imagination' (Coeckelbergh, 2006). In my case, what grew both this reflection on the ethics of flood risk management in the UK and my own moral imagination was the work that we undertook through the Ryedale Flood Research Group. As a member of that group, I was forced to dissociate from my normal institutional networks, and to rebuild networks with others (Landström *et al.*, in press). Ironically, this is no different to what I, as a scientist, do when I move into a new research area or choose to pursue a new research method. The difference was that my re-association was not with a different community of scientists but with those who have to live with flood risk. Ultimately, my experience is a critique of the growth of institutions as not just intermediaries but as knowledge gatekeepers, where those gatekeepers commission scientific knowledge on the one hand, regulate who produces that knowledge and how it is produced, and communicate it on the other. Deliberation and consultation is an inadequate response to the isolation of scientific knowledge from lay experiences that follows. Not only is the framing of institutional knowledge, as geographically extensive and comparative, hard to reconcile with the kind of framing that comes from those who live with risk, but if knowledge is to be paramount in environmental decision-making then the imperfect nature of that knowledge needs also to be experienced. A failure to do so will simply lead to a breakdown of trust, especially during catastrophic events, in those intermediaries necessary to sustain difficult decisions. It becomes reinforced by ethical questions regarding precisely whose knowledge counts in making those decisions.

The challenge is that co-production of knowledge is hard. Seeing a public as in need of education, and then communicating the associated knowledge is surprisingly easy: communication can become a discrete act once the scientific account has been deemed as acceptable by an intermediary. Even consultation and debate is straightforward as it can be reduced to a discrete number of interventions. Co-producing knowledge obviates the need for communication, even consultation, but it has to be replaced by a continual engagement throughout the risk management process; one that recognises that the means by which a decision is made can become more important than the eventual decision taken itself. If co-production can be achieved; and generating, managing and acting upon knowledge can become generally redistributed; then a genuine result could be a much deeper individual and collective acceptance that living with risk means living with risk, rather than being forced to construct events as extremes that have exceeded all expectations (Rozario, 2007).

Acknowledgements

This chapter is partly based upon work funded by grant RES-227-250-018 from the Rural Economy and Land Use programme of three UK research councils (BBSRC, ESRC and NERC) and DEFRA, awarded to Professor Sarah Whatmore, Professor Neil Ward and myself. I am particularly grateful to both the academic and local members of the Ryedale Flood Research Group, to Gillian Willis (Oxford University) and to those others who agreed to be interviewed for this research.

References

Adams, J. (1995) *Risk*, UCL Press, London, UK.

Bradbrook, K.F., Lane, S.N., Waller, S.G., Bates, P.D. (2004) Two dimensional diffusion wave modelling of flood inundation using a simplified channel representation. *International Journal of River Basin Management*, **3**, 1–13.

Callon, M. (1999) The role of lay people in the production and dissemination of scientific knowledge. *Science Technology and Human Values*, **4**, 81–94.

Cockerill, K., Passell, H. and Tidwell, V. (2006) Cooperative modelling building bridges between science and the public. *Journal of the American Water Resources Association*, **42**, 457–471.

Coeckelbergh, M. (2006) Regulation or responsibility? Autonomy, moral imagination, and engineering. *Science, Technology and Human Values*, **31**, 237–260.

Collins, H.M. and Evans, R. (2002) The Third Wave of Science Studies: Studies of Expertise and Experience. *Social Studies of Science*, **32**, 235–296.

Costanza, R. and Ruth, M. (1998) Using dynamic modeling to scope environmental problems and build consensus. *Environmental Management*, **22**, 183–195.

DEFRA (2004) *Community and public participation: risk communication and improving decision-making in flood and coastal defence*. Research and Development Technical Report FD2007/TR, Department of the Environment, Food and Rural Affairs, London.

DEFRA (2008) *Social Justice in the Context of Flood and Coastal Erosion Risk Management: A Review of Policy and Practice*. DEFRA R&D Technical Report FD2605/TR, London.

Environment Agency (2004a) *Building Trust with Communities: A Background Report for Environment Agency Staff*, Internal Publication.

Environment Agency (2004b) *Building Trust with Communities: A Toolkit for Staff*, Internal Publication.

Environment Agency (2005) *Improving community and citizen engagement in flood risk management decision making, delivery and flood response*. Environment Agency R&D Technical Report SC040033/SR3, Environment Agency, Bristol.

Funtowicz, S.O. and Ravetz, J.R. (1991) A new scientific methodology for global environmental issues, in *Ecological Economics* (ed R. Costanza), Columbia University Press, New York, pp. 137–152.

Funtowicz, S.O. and Ravetz, J.R. (1993) Science for the post-normal age. *Futures*, **25**, 740–755.

Hale, E.O. (1993) Successful public involvement. *Journal of Environmental Health*, **55**, 17–19.

Harrison, S. and Mort, M. (1998) Which champions, which people? Public and user involvement in health care as a technology of legitimation. *Social Policy and Administration*, **32**, 60–70.

Jasanoff, S. (2003) Breaking the Waves in Science Studies. *Social Studies of Science*, **33**, 389–400.

Johnson, C., Penning-Rowsell, E. and Parker D. (2007) Natural and imposed injustices: the challenges in implementing 'fair' flood risk management policy in England. *Geographical Journal*, **173**, 374–390.

Klauser, F.R. and Ruegg, J. (this volume) Finding the Right Balance: Interacting Security and Business Concerns at Geneva International Airport. In (eds M.B. Keanes, F.O. Klauser and S.N. Lane).

Landström, C., Whatmore, S.J. and Lane, S.N. (2011) Virtual Engineering: computer simulation modelling for flood risk management in England and Wales. *Science Studies* Forthcoming.

Landström, C., Whatmore, S.J., Lane, S.N., Odoni, N., Ward, N. and Bradley, S. (in press) Dissociations and attachments enabling scientists to co-produce flood risk knowledge with affected residents. Forthcoming in *Environment and Planning A*.

Lane, S.N., Landstrom, C. and Whatmore, S.J. (2011a) Imagining flood futures: risk assessment and management in practice. *Philosophical Transactions of the Royal Society, A*, **369**, 1784–1806.

Lane, S. N., Odoni, N., Landström, C., Whatmore, S. J., Ward, N. and Bradley, S. (2011) Doing flood risk science differently: an experiment in radical scientific method. *Transactions of the Institute of British Geographers*, **36**, 15–36.

Lane, S.N. (2011) Making mathematical models perform in geographical space(s), in *Handbook of Geographical Knowledge* (eds J. Agnew and D. Livingstone), Sage, London.

Latour, B. (2004) Why has critique run out of steam? From matters of fact to matters of concern. *Critical Inquiry*, **30**(2), 225–248.

Mustafa, D. (2005) The production of an urban hazardscape in Pakistan: modernity, vulnerability, and the range of choice. *Annals of the Association of American Geographers*, **95**, 566–586.

Ohl, C.A. and Tapsell, S. (2000) Flooding and human health: The dangers posed are not always obvious. *British Medical Journal*, **321**, 1178.

Rozario, K. (2007) *The Culture of Calamity: Disaster and the Making of Modern America*, University of Chicago Press, Chicago.

Slovic, P. (2000) *Perception of risk*, Earthscan, London.

Stirling, A. (2008) 'Opening Up' and 'Closing Down': power, participation, and pluralism in the social appraisal of technology. *Science, Technology and Human Values*, **33**, 262–294.

Sturgis, P. and Allum, N. (2002) Science in society: re-evaluating the deficit model of public attitudes. *Public Understanding of Science*, **13**, 55–74.

White, G. F. (1942) *Human adjustment to floods: a geographical approach to the flood problem in the United States,* Department of Geography Research Paper 29, University of Chicago.

Wynne, B. (2003) Seasick on the Third Wave? Subverting the Hegemony of Propositionalism, *Social Studies of Science,* **33**, 401–17.

Yearley, S. (2006) Bridging the science-policy divide in urban air-quality management: evaluating ways to make models more robust through public engagement. *Environment and Planning C,* **24**, 701–714.

Yu, D. and Lane, S.N. (2006) Urban fluvial flood modelling using a two-dimensional diffusion wave treatment: 1 Mesh resolution effects. *Hydrological Processes,* **20**, 1541–65.

Yu, D. and Lane, S.N. (2006) Urban fluvial flood modelling using a two-dimensional diffusion wave treatment: 1 Development of a sub-grid scale treatment. *Hydrological Processes,* **20**, 1567–1583.

CHAPTER 9

In the Wake of the Tsunami: Researching Across Disciplines and Developmental Spaces in Southern Thailand

Jonathan Rigg[1], *Lisa Law*[2], *May Tan-Mullins*[3], *Carl Grundy-Warr*[4] *& Benjamin Horton*[5]

[1]Department of Geography, Durham University, Durham, UK
[2]School of Earth and Environmental Sciences, James Cook University, Australia
[3]Division of International Studies, University of Nottingham, Ningbo, China
[4]Department of Geography, National University of Singapore, Singapore
[5]Department of Earth and Environmental Science, University of Pennsylvania, Pennsylvania, USA

Introduction

In this chapter, we reflect on our experiences undertaking risk research as part of a multi-national, multi-disciplinary team of scholars and researchers exploring the impacts and responses to the Boxing Day 2004 Indian Ocean tsunami in Thailand. The original, exploratory research was funded by the US National Science Foundation and sought to investigate how local people (households and individuals), groups (NGOs and civil society organisations), and government agencies responded to the tsunami, and the patterns of impact, resilience and recovery that ensued (see below for further details). In this chapter, however, we focus not on these research issues, which we have explored in a number of other papers,[1] but on three wider sets of debates that have relevance to the study and understanding of 'risk' more generally.

---

[1] We have published a number of papers based on the fieldwork. See Rigg *et al.* 2005 and 2008; Tan-Mullins, Rigg and Grundy-Warr, 2008; Tan-Mullins *et al.* 2007; Buranakul *et al.* 2005; and Horton *et al.* 2008. The last named author of this chapter was the PI on the project and is a natural scientist; the remainder of us are social scientists.

Critical Risk Research: Practices, Politics and Ethics, First Edition.
Edited by Matthew Kearnes, Francisco Klauser and Stuart Lane.

The first centres on the experience of working in an interdisciplinary team, and particularly on a topic which was so acutely sensitive at the time we carried out the fieldwork (barely six months after the event). Thousands of people died in the areas of southern Thailand where we were working, and researching in this context was often moving, and sometimes quite harrowing. Issues that are germane to this first set of debates include our approach to 'the field' in general and to our research subjects in particular. The second set of debates relates to how risk is conceptualised and compartmentalised across the social and natural sciences, and by researchers and different populations of research subjects. In particular we consider the question of what 'counts' as evidence across disciplines. This then leads to a third theme that the chapter addresses, namely the spatiality of risk, vulnerability and opportunity and where we draw the lines of enquiry around a natural event such as a tsunami or, for that matter, an earthquake, cyclone or landslide. All these three themes are then drawn together in a wider consideration of the disciplinary boundaries that shape and colour research, and the challenge and difficulty of transcending such boundaries. In summary, these themes enable us to explore the question of how we 'do' risk research. This involves a set of ethical concerns pertaining to our conduct in the field; a series of methodological issues related to where the boundaries of risk research lie and where we should look to collect evidence; and a set of contextual challenges linked to the 'geography' and therefore spatiality of risk.

Setting the scene

The Tsunami event

On the morning of 26 December 2004, the Aceh-Andaman earthquake (Mw $\sim$ 9.2) occurred in the Indian Ocean, 160 km west of the Indonesian island of Sumatra. In the last 100 years only one earthquake has been more powerful: the Great Chilean Earthquake of 1960 (Mw $\sim$ 9.5). The earthquake deformed the ocean floor and generated a tsunami which was the cause of one of the greatest natural disasters of the last century. Around 250 000 people died in an arc of devastation from Indonesia through Malaysia, Thailand, India, Sri Lanka and East Africa. The tsunami, perhaps because of a combination of the date (over Christmas), the striking television coverage, and the number of foreign tourists who were affected held the headlines for a strikingly long period of time. It also generated an unprecedented wave of generosity from the public in the Global North and from many governments (see Brown and Minty, 2006).[2] According to

[2] Although not all the assistance pledged was delivered.

Table 9.1 Casualty data for Thailand

Province	Deceased				Injured			Missing		
	Thai	Non-Thai	No info	Total	Thai	Non-Thai	Total	Thai	Non-Thai	Total
Phangnga	1,266	1,633	1,325	4,224	4,344	1,253	5,597	1,428	305	1,733
Krabi	357	203	161	721	808	568	1,376	329	240	569
Phuket	151	111	17	279	591	520	1,111	256	364	620
Ranong	156	4	0	160	215	31	246	9	0	9
Satun	6	0	0	6	15	0	15	0	0	0
Trang	3	2	0	5	92	20	112	1	0	1
TOTAL	**1,939**	**1,953**	**1,503**	**5,395**	**6,065**	**2,392**	**8,457**	**2,023**	**909**	**2,932**

Note: casualty data as of 24th March 2005.

Thomas and Fritz 'the world responded by donating more than $13 billion and initiating the largest relief effort in history' (2006: 114).

Large areas of coastal southern Thailand – as well as Indonesia, Sri Lanka and India – were transformed when the tsunami swept ashore. Maximum tsunami wave heights were in excess of 34 metres in Sumatra (Tsuji, 2005). Officially, there were 5395 confirmed deaths in Thailand, with a further 2932 people listed as missing. Most of the deaths were concentrated in the provinces of Phang-nga and Krabi, with smaller numbers of fatalities in Phuket, Ranong, Satun and Trang (Table 9.1). The large number of non-Thai fatalities in part reflects the fact that these provinces of southern Thailand are popular international tourist destinations, and Christmas is a peak in the tourism calendar. Although the figures give the impression of great accuracy, in all likelihood the true toll was significantly higher because the areas affected also supported a large population of unregistered migrants, mainly from Burma (Myanmar), working in the tourism, construction and fisheries industries. Given their unregistered and therefore illegal status in Thailand coupled with social and political conditions in Burma, it is highly likely that many Burmese who died in the tsunami were never registered as missing. We can assume that the large number of unidentified deceased (see Table 9.1) mainly comprise these migrants from Burma, who have been termed the 'invisible' victims of the disaster (Hakoda, 2005; and see Yang, 2007).

Research sites and methods

In February 2005 a team led by Dr Benjamin Horton of the University of Pennsylvania was awarded an SGER grant by the National Science Foundation ('Indian Ocean tsunamis: socio-economic impacts on the Malay-Thai peninsula') to undertake exploratory research on the tsunami in

Malaysia and Thailand. The proposal submitted to the NSF set out to investigate the following themes:

- Local people's understanding and interpretation of the tsunami as an 'event'.
- The differing patterns of resilience and recovery among affected households and communities.
- The response patterns of local communities, and government and non-governmental agencies.
- The role of 'social capital' – articulated at the local, community level and embedded in the traditions of community life – in resilience and recovery.

The fieldwork in Thailand centred on three main sites: Koh Lanta, Koh Phi Phi and Khao Lak (Figure 9.1). We also made brief visits to several other locations in this area of southern Thailand, as well as an initial pilot visit to Penang and Langkawi in Malaysia. We chose Koh Phi Phi as a small, tourist (backpacker) oriented island economy with high levels of damage and casualties; Koh Lanta as a site supporting a significant population of fisherfolk with a long presence in the area; and Khao Lak as a mainland site with the highest number of casualties in Thailand and with a mixed tourism-fishing economy, and a large number of unregistered migrants. The specific sites where we chose to undertake fieldwork were also mixed in terms of their ethnic, social and religious make-up, including Buddhist and Muslim villages, tourist enclaves and areas with large numbers of Burmese migrant labourers. We decided not to visit the island of Phuket because this site was, at the time, already the subject of considerable attention by other researchers.

Khao Lak was the area of Thailand most severely hit by the tsunami. The official death toll in March 2005 for the province of Phang-nga (of which Khao Lak is a part) was 4224, with another 1733 people listed as missing. These comprise the large majority of the fatalities in Thailand. Local people, however, found these figures hard to reconcile with the scale of the destruction they had witnessed and, in consequence, believed the real death toll to be considerably higher. The fishing villages at the northern end of the Khao Lak strip such as Ban Nam Khem and Ban Laem Pom had significant populations of migrants living there, including many unregistered Burmese workers (see above).[3]

---

[3] Before the tsunami there were 22 000 registered Burmese workers in Phang-nga province, most of them employed in Takua Pa district (where Khao Lak is located) in the fishing and construction industries. However the number of legal and illegal Burmese workers in Thailand probably numbered more than two million (http://www.achr.net/000ACHRTsunami/Burma%20TS/Tsunami%20Burma.htm).

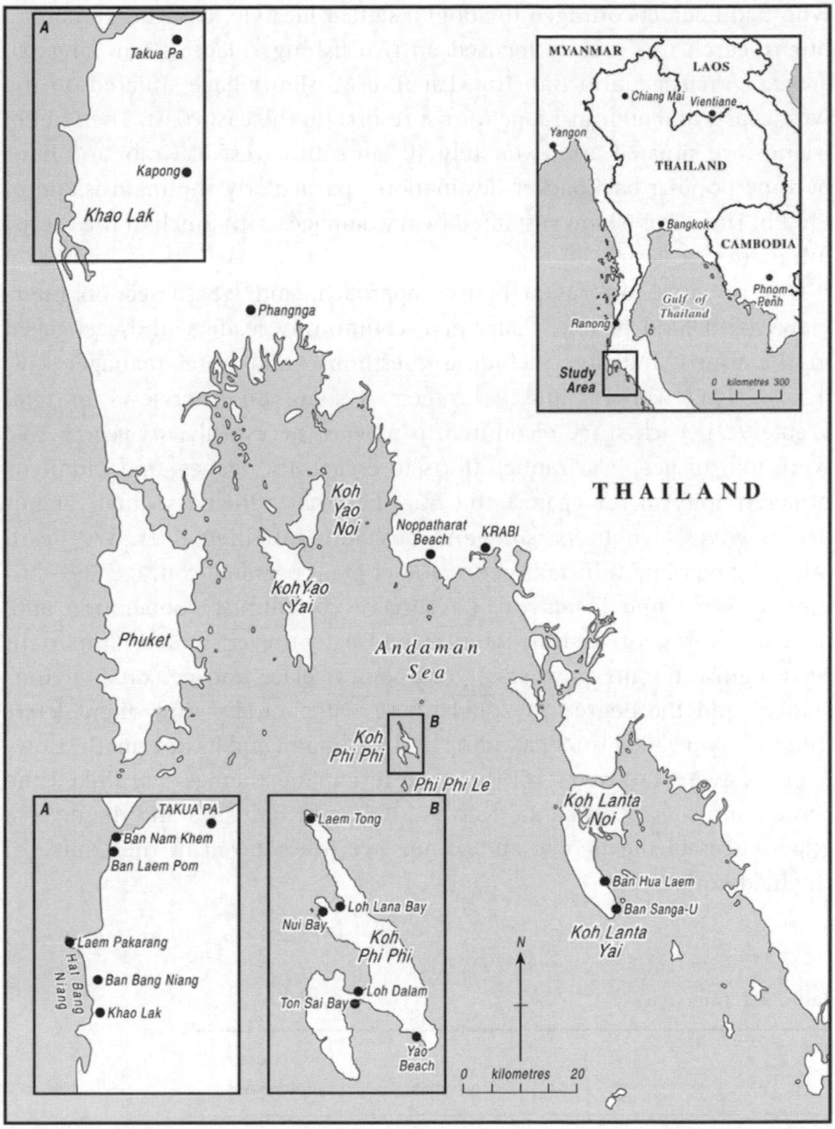

Figure 9.1 The tsunami field sites in southern Thailand *Source:* Rigg *et al.* 2008

The province of Krabi, which was the second most severely affected province in Thailand after Phang-nga with 721 fatalities, 1376 injured and 569 missing, covered our other two main research sites: the islands of Koh Lanta Yai and Koh Phi Phi (Figure 9.1). Koh Lanta Yai had a concentration of tourist resorts and guesthouses on the east coast, as well as eight fishing villages. The populations of these villages were largely Muslim (as opposed to Theravada Buddhist), and some were sea gypsy (or *chao lay*) communities – formerly nomadic boat people of Indo-Malay origin who had been encouraged to adopt a settled lifestyle. On Koh Lanta Yai, our research was mainly focused on two fishing villages (Ban Sanga-U, a *chao lay* village; and Ban Hua Laem, a Muslim village situated on the west coast) and on mid-range tourist resorts on the east coast. The Phi Phi Islands are situated approximately 40 km south-west of Krabi and have become popular backpacker destinations, particularly the main island of Phi Phi Don. This island sustained heavy damage, with much of the tourist infrastructure destroyed.

Our research took a qualitative approach, and was based on interviews with local people, fishermen, community leaders, those engaged in the tourist industry including guesthouse and hotel managers, officials, NGO workers and volunteers – some 80 interviews in total (Table 9.2). Each story recounted to us was, necessarily, *sui generis*. We were told of people's frantic efforts to escape the waves, and moments of serendipity to set against the tragedies that afflicted so many families as wives, daughters, sons and husbands lost their lives. We heard tales of government intransigence and of great personal courage and generosity. We found instances of resurgent community cooperation and, in a few cases, of dysfunctional response to the crisis. It is important to recognise the uniqueness of geographical place and personal circumstances and the degree to which these will mould – and often determine – any detailed understanding of the tsunami and its aftermath. However we were also aware of the need to explore themes that linked the three main sites (Koh Lanta, Koh Phi Phi and Khao Lak) and the diverse experiences of the people and groups we interviewed in the course of the fieldwork.

Table 9.2 Interviews

	Interviews	Semi-structured interviews
Tourism employees	6	7
Hotel managers and restaurant owners	5	2
Fishers	8	13
NGO workers	3	0
Volunteers	12	0
Village leaders	2	0
Villagers	8	12
Others	2	1
TOTAL	**46**	**35**

Researching across disciplines

'Evidence': what is it, where is it situated and how is it valued?

A recurring feature of the collaborations that were needed to undertake this research were the different ways in which the natural science and social science members of the research team looked for 'evidence'. This included what we thought 'counted' as evidence; where we considered it might be located; how we measured, recorded and valued it; and how this has then shaped the published research outputs.

This difference in philosophy first became evident during our first, preliminary reconnaissance trip to Penang, Malaysia.[4] As part of this meeting we visited a number of coastal sites on the east coast of Penang that had been affected by the tsunami. When we left the minibus on one of these forays to the field, the natural scientists walked down to the shoreline to look for physical evidence of the surge in the form of debris or damage, its size, distribution and orientation; at the same time, the social scientists made their way to a pier where a group of locals were sitting under an awning (Figure 9.2). This happened spontaneously, without prior planning and discussion. We – the social scientists – asked the men on the pier whether they experienced the tsunami (as Malay fishermen, they had), what form the tsunami had taken, the time of the waves' arrival, direction, wave height, the number of tsunami waves, and so forth. We were also pointed to water level marks on coastal structures that revealed the height and variability of the tsunami waves (Figure 9.2), and to beached fishing boats that had been carried inland, through coconut groves, by the force of the wave. The natural scientists, meanwhile, surveyed the coastline looking for physical evidence of the surge. In retrospect, the natural scientists considered this oral, interview-based information very important for the interpretation of the tsunami deposits and a considerable advance in terms of approach, improving accuracy and directing attention to evidence that might otherwise have been overlooked. At the time however, our responses in the field were notably different, informed by the nature of the data (evidence) for which we were in search.

At the time this struck us as reflecting a very different mindset as to where research of this type starts. For the social scientists, it was in eyewitness accounts, in 'anecdotal' and qualitative evidence gleaned from chance meetings and interviews. We were intent on constructing multiple stories

---

[4] Prior to the main period of fieldwork in southern Thailand, many of the research team met in advance in Penang and Langkawi to discuss methods and to carry out pilot work.

Figure 9.2 Interviewing Malay fishermen in Penang, who had experienced the tsunami first hand (May 2005). The height of the surge was indicated to us as the painted marks around 0.5 m above the man's head on the metal pole on the right hand side of the image

of the tsunami derived from the mixed experiences of those present at the time, recognising that these would not necessarily chime. It was exploratory, open-ended, inductive and grounded theoretically. The natural scientists, meanwhile, were intent on finding hard evidence in the form of measureable levels of debris and damage, linked to a working hypothesis regarding the signature of the event as recorded in the physical debris. The witnesses to the event, for us, were the people who were there; the witnesses to the event for our natural science co-workers were waiting to be unearthed, quite literally, in the landscape.

A research field guide to qualitative methods states: 'Qualitative research is especially effective in obtaining culturally specific information about the values, opinions, behaviors, and social contexts of particular populations' (FHI, n.d.). It was this research and field philosophy which shaped our research in southern Thailand, our engagement with the research subjects, and which led us to adopt a broadly qualitative epistemology. We took the decision to undertake our research largely through purposively selected interviews (see Figure 9.3) because we wished to understand, in as much detail as possible, people's experiences of the tsunami and their reactions to it. We rejected undertaking a sample survey because we thought this would lead to a narrow and thin investigation dictated by certain prior

Figure 9.3 May Tan-Mullins interviewing a Muslim woman and member of a fishing household on Koh Phi Phi (July 2005)

assumptions as to the pertinent issues; we wished to keep the research, particularly at the start, as 'open' as possible, given its exploratory nature. We also felt that such an approach, with clipboards and photocopied survey questionnaire forms, would be inappropriate in the circumstances. Indeed, on many occasions we found ourselves drawn into long and often highly personal discussions as our interviewees talked about lost loved ones and ruined lives.[5] We thought that ethically we should permit the subjects of the research to speak for themselves and set their own agendas rather than having these dictated *a priori* (i.e. deductively). The research was, therefore, inductive and exploratory. There was a series of questions that we wished to illuminate about *why* people acted and behaved as they did during and following the tsunami, and a qualitative approach was deemed the most appropriate.

What has become clear, however, is that notwithstanding the growing and widespread use of qualitative methods in the social sciences, there remains the view in some quarters that such methods are unscientific, with the data and information produced often pejoratively labelled 'anecdotal'

[5] This presented an ethical dilemma of another sort: were we suitably trained to act as *de facto* counsellors? It was sometimes distinctly unsettling to be placed in this position, and all of us felt inadequate to the task.

or 'journalistic'. This became evident in one referee's responses to a paper based on the research discussed here that we submitted to a peer review journal with a 'disasters' remit (see Table 9.3). Referee #1 considered, it seems, interview-based research of the type we employed to be necessarily anecdotal and journalistic, and instead recommended the survey of 'a large, homogeneous sample of people'. The qualitative interviews that we undertook were dismissed as 'opportunistic' and 'not science'. Referee #2, on the other hand, found the approach and the paper an 'interesting and welcome addition', and the case studies 'rich, illuminating, sensitively written and in many ways relevant'. Referee #1 concluded his review with the dismissive '...given further field work... [this] could blossom into something of value'. The point of this is not to suggest that the paper *should* have been accepted for publication by the journal concerned, but to highlight the chasm that exists in what 'counts' as legitimate evidence, indeed what counts as science.[6]

The research on which this chapter is based was framed as an inter-disciplinary study, and justified to the funding body as such. We felt that understanding events such as the tsunami, like so much research which engages with 'real' world problems from health threats to climate change adaptation, requires interdisciplinary collaboration, drawing on the expertise of several subject areas, often in a multi-level way (in other words, which crosses fields, such as the environmental, behavioural and social [see Anderson, 1999]). And yet we found it a continuing struggle to escape and work beyond the confines of our disciplinary instincts. It is notable that we have not published together in a manner that could be described as 'inter-disciplinary', by which we mean research that involves the combination or integration of ideas, data, techniques, concepts and methods across at least two disciplines (see Table 9.3). This was despite the fact that we collaborated on shaping the original application, took part in a shared period of pilot work in Malaysia, and lived and worked closely together in southern Thailand during the main period of fieldwork. This highlights the difficulty of moving beyond the rhetoric of inter-disciplinarity; our published outputs have tended to remain stubbornly segmented by disciplinary field and, at best, are multi-disciplinary (Table 9.4). This was, no doubt, partly because of the challenge of integrating methods across the social and natural sciences in a coherent, rather than just an opportunistic manner. It also reflects, however, the fact that most journals, and particularly most high impact journals, are resolutely disciplinary. As Brewer memorably put it, 'The world has problems, but universities have departments' (1999: 328).

---

[6] The revised paper has since been published – see Rigg *et al.* 2008.

Table 9.3 Valuing evidence

Referee #1	Referee #2
The *language is overly academic and obfuscates the main messages* in the paper. This mss is full to the *brim with the jargon of post-modern, "critical" geography – a small and obscure, mostly American academic enterprise.*	This is an impressive and enlightening paper. It is particularly well written and edited. The 'tsunami footprint' conclusion is an *interesting and welcome addition* to analyses regarding the impact on and role of diasporas in disasters.
The analysis interweaves the authors' local findings with secondary sources, and positions this within a wider literature . . . *it has more in common with journalism than solid research*. . . . had the authors been able to interview a *large, homogeneous sample of people* . . . then they might have been able to say something about tsunami recovery and poverty. But one month's field work with no follow up, and with a *rather opportunistically chosen*, small sample of informants, they *cannot make credible statements* on this important subject.	All-in-all, this is a very good paper and congratulations to all concerned.
[The paper] leans heavily on some individual "stories". This and the prior section might be *interesting journalism* – of the kind one sees in the press releases of international NGOs and in the blogs of Reuters AlertNet. . . . However, *this is not science*.	The three mini-case studies are quite central to the 'mapping and explaining the trajectories of recovery'. They are *rich, illuminating, sensitively written* and in many ways relevant.
Most seriously, there is *no consideration of the practical and policy implications* of the findings . . . *Given further field work . . . this work could blossom into something of value.*	Again, this is a very good paper . . . it will be an excellent contribution.

This interpretation of the challenges we faced in our tsunami research is relatively benign, and implies that we have to work harder at understanding each other, and get into the habit of working in multidisciplinary teams so that inter-disciplinary scholarship becomes normalised. In this way, disciplines and their constitutive 'frames of reference, methodological approaches, topics, theoretical canons and technologies' (Cronin, 2008: 3) can be opened up. This in itself is a considerable challenge because so much scholarship is based on and ordered around disciplinary divides, from the administration of universities (in departments), to journals, curricula and degrees, and funding bodies which are all, frequently, based on a disciplinary ordering. But it is possible to see the problem as possibly even more intractable than this. In an influential paper, Gieryn (1983) sees the divide between science and non-science as actively cultivated by scientists as a means to preserve and protect their intellectual authority, to prevent

Table 9.4 Published papers emerging from the research

Social Science	Multi-disciplinary	Natural Science
Rigg, Jonathan, Lisa Law, May Tan-Mullins, Carl Grundy-Warr (2005) "The Indian Ocean tsunami: socio-economic impacts in Thailand", *The Geographical Journal* 171(4): 374–79.	Bird, M., Cowie, S., Eong Ong, J., Hawkes, A., Horton, B.P., Tan Shau Hwai, A., Wooi Khoon, G., Macgregor, C., Tiong Sa, T. and Yasin, Z. (2007)"Indian Ocean tsunamis: environmental and socio-economic impacts in Langkawi, Malaysia", *The Geographical Journal* 173: 103–117. (Natural science leaning)	Hawkes, A. D., Bird, M., Cowie, S., Grundy-Warr, C., Horton, B.P., Tan Shau Hwai, A., Law, L., Macgregor, C., Nott, J., Eong Ong, J., Rigg, J., Robinson, R., Tan-Mullins, M., Tiong Sa, T. and Zulfigar, Y. (2007) "The sediments deposited by the 2004 Indian Ocean tsunami along the Malaysia-Thailand peninsula", *Marine Geology* 242: 169–190.
Buranakul, S., Grundy-Warr, K., Horton, B., Law, L., Rigg, J. and Tan-Mullins, M. (2005) "The Asian tsunami, academics and academic research", *Singapore Journal of Tropical Geography* 26(2): 244–48.	Horton, B.P., Bird, M., Birkland, T., Cowie, S., Grundy-Warr, C., Hawkes, A.D., Tan Shau Hwai, A., Law, L., Macgregor, C., Nott, J., Eong Ong, J., Rigg, J., Robinson, R., Tan-Mullins, M., Tiong Sa, T. and Zulfigar, Y. (2008) "Environmental and socio-economic dynamics of the Indian Ocean Tsunami in Penang, Malaysia", *Singapore Journal of Tropical Geography* 29: 307–324. (Social science leaning)	
Tan-Mullins, May, Jonathan Rigg, Lisa Law and Carl Grundy-Warr (2007) "Mapping the local politics of aid: structures, networks and capitals in post-tsunami Thailand", *Progress in Development Studies* 7(4): 327–344.		

Table 9.4 (*Continued*)

Social Science	Multi-disciplinary	Natural Science
May Tan-Mullins, Jonathan Rigg and Carl Grundy-Warr (2008) "Responses and resilience of fisherfolks on the tsunami event in Southern Thailand", in Tim Doyle and Melissa Risely (edits) *Crucible for survivial: environmental security and justice in the Indian Ocean region*, New Jersey, New York and London: Rutgers University Press, pp. 116–129.		
Rigg, Jonathan, May Tan-Mullins, Lisa Law and Carl Grundy-Warr (2008) "Grounding a natural disaster: Thailand and the 2004 tsunami", *Asia Pacific Viewpoint* 49(2): 137–154.		

incursion into their territory from non-(or 'pseudo'-)scientists, and to pursue material resources (i.e. funding). The divide, therefore, is not a philosophical one, nor one that might be bridged by normalising engagement, but one pursued for strategic reasons.

Geographies of fear, psychologies of explanation

'Natural' disasters such as the 2004 tsunami characteristically bring together a sometimes uncomfortable mélange of actors, explanatory cultures, agencies, world views and organisations. The event created a requirement, of course, for humanitarian relief; it also shaped a need for succour and for explanation. Into this were drawn religious groups, NGOs, local community leaders, officials, politicians – and scientists (academics). The nature and unprecedented scale of the tsunami instilled a sense of fear; of the unknown, the uncertain and the uncontrollable. Victims sought to come to terms with the event, attempting to explain and rationalise the *khluen yak* (คลื่นยักษ์), or giant wave. This rationalisation involved an engagement, on the part of the victims, with both the science of earthquakes and tsunamis, and the incorporation/embedding of this science within traditional Thai beliefs, and within Buddhism and Islam. It was striking how many times we were asked questions of a scientific nature, and Thai language newspapers provided full scientific explanations of the tsunami, along with supporting diagrams. But our interviewees' interest in 'science' did not displace the value and importance they attached to other modes of explanation, religious and 'traditional'.

For residents in the immediately affected coastal areas, the day the tsunami washed ashore was the moment when the normal routines of life were swept away: the wave came to represent a 'break' (taek, แตก) in their lives and in community history. Many lost members of their families and, because so many bodies were never recovered, coming to terms with their losses became even more difficult and traumatic. For those who lived and made a living by and on the sea (including both fisherfolk and those in the tourist industry), the coast was transformed from a site of sustenance, to a place of loss and of fear.

A cultural 'trait' of the Thai people – and this applies to both Buddhists and Muslims – is a well-honed belief in ghosts, spirits and spirit mediums (see Mills, 1995; Piker, 1968; and Pattana Kitiarsa, 2005).[7] While more than four decades may have passed since Piker wrote his paper on the belief systems of northern Thai peasants, his distillation continues to resonate, even in an increasingly modernist and urbanised Thailand: 'Theravada Buddhism notwithstanding . . . [the Thai peasant] also propitiates spirits, practices magic and divination, and fears ghosts and sorcerers' (Piker, 1968: 384). Spirit houses are common as Thais believe that wandering spirits require shelter and a place to reside if they are to be happy and satisfied; otherwise they may haunt the houses of the living. More particularly, there is a belief that wandering human spirits are the spirits of those who have died of unknown causes, or who have died suddenly.[8] Because the tsunami was beyond the experience of most Thais it was translated, according to traditional belief, into thousands of wandering spirits in the affected areas. One of our informants on Koh Phi Phi, Daa, explained to us:

> All my friends died with their eyes open, as they did not know what killed them. This is not good in Thai society, as they will turn into wandering spirits.

The belief that they might be 'disturbed by ghosts' kept many domestic tourists away from those areas where people died.

---

[7] The large majority of Thais are Theravada Buddhist. However coastal areas in the South also have significant Muslim populations.

[8] One debate on Koh Phi Phi while we were there was how the victims should be commemorated. Many *farang* (foreigners) supported the construction of a prominent, physical memorial to those who had lost their lives in the tsunami; local Thais preferred rites – the chanting of monks to bring peace to the spirits of the dead – and many were not keen on a memorial.

Explanations for the tsunami in the aftermath of the event often combined science, religion and, more broadly, 'cultural' interpretations. The Thai media played a role in educating the public that the tsunami was generated by an undersea earthquake in the Indian Ocean; some informants went into considerable detail explaining to us how the seabed was deformed, even those such as fisherfolk with comparatively limited formal education. But to set against this were the views of informants, generally more elderly and often Muslim, who told us that the tsunami was 'God's way of telling us that we have sinned'. On Nopporathara Beach in Krabi province, one informant said that it was because younger Muslims had deviated from Islam, consuming alcohol and drugs. As retribution, God sent the tsunami to cleanse the area of sin. On Koh Phi Phi, where many local people are also Muslim, the invasion of foreign tourists drinking heavily, eating pork, dressing inappropriately and sunbathing topless was also seen by some as an explanation for the tsunami. The tsunami was, in other words, seen to be propelled by human actions and responsibilities.

As, in the evenings, we jointly discussed the explanations that were put to us for the tsunami, we debated their compartmentalisation into 'science' and 'non-science'. Those informants who explained and understood (and therefore tried to come to terms with) the tsunami in religious or supernatural ways also allied – and often quite seamlessly – the supernatural/religious with the scientific. For these people, the tsunami was both a natural event *and* a religious/supernatural occurrence. Human actions, supernatural forces and tectonic processes formed a triptych of explanation. Some studies have suggested that a belief in spirits in Thailand (and elsewhere) is seen as part of a counter-hegemonic discourse, one that seeks to challenge the state-orchestrated modernist mode of explanation (see Mills, 1995 for example). On the basis of our research following the tsunami, we identified no such confrontation where modernist (scientific) and traditional explanations provided alternative interpretations; rather the two nested together. Nonetheless, wrinkles in this apparent melding of the scientific and the religious, or rationality with fatalism, did sometimes make themselves evident. In the Muslim fishing settlement of Ban Hua Laem on Koh Lanta, there seemed to be three broad responses by local people in the months after the tsunami. First of all there was a reinvigoration of religious belief, with the village mosque and Muslim association becoming especially active. Second there was a degree of fatalism among some inhabitants – 'if it happens (again), it happens', as one respondent put it. A Muslim informant explained that in the past he had prayed to God for protection while at sea through performing rituals each year. But he openly wondered whether the ceremony was sufficient in these new, post-tsunami times: 'We do the same as before but don't feel good about it, as it is no longer the same as before'. The tsunami was being seen as a watershed by many, dividing what use to be (whether in economic, cultural or psychological terms), from what had become. There were also, and third, those who favoured a more practical community response. In a meeting between the local headman and several village leaders to discuss and vote upon the use of some remaining aid money, the discussion centred on whether the money should be used for a village early-warning system or a new football pitch. After much deliberation and argument, the early-warning system narrowly won the day.

As Merli (2010) argues in her paper, based on research in the Muslim-majority province of Satun in southern Thailand, local discourses of explanation for the tsunami were mainly religious, but were also plural, employing multiple theodicies linked to the syncretic nature of religious belief in the area, and most people attributed the disaster to human responsibility:

> People weave together different theodicies.... In presenting their views, Muslims and Buddhists reflect on their respective theodicies, and in pointing to the similarities and differences they define who they are with respect to the other, setting social boundaries. These kinds of reflections are the result of the local composition of the population, the territory they share and their history.... Shunning a univocal causal explanation of the tsunami as a natural disaster, Muslims' attempt to combine scientific and theological explanations serves the aims of disciplining the self and the other, by stressing the ultimate human responsibility for such events. There is thus an attempt to reconcile the other-worldly and the this worldly, to create a new theodicy (Merli, 2010: 110).

Fear manifested itself in various ways. Many former fishermen abandoned the sea and turned to other occupations. This occurred at the same time as aid was being channelled into re-equipping fishing households with new boats to replace those they had lost. While the focus of assistance was on 're-building' (i.e. re-creating) livelihoods, the concern of those affected was not infrequently concentrated on *transforming* livelihoods (see Lebel, Supaporn Khrutmuang and Manuta, 2006). A similar distinction between re-creation and trans-formation was evident in the villages themselves. In Ban Nam Khem, a Buddhist fishing community in Khao Lak which was virtually wiped out by the tsunami, 80 households continued to live, some six months after the event, at Wat (Monastery) Nam Khem, inland from the coast and at a higher elevation than the village. By that time, the village has been rebuilt, on the same site as the original settlement, but families were reluctant to move to their new houses, instead sleeping in the monastery at night and occupying their new village during the day (see Figure 9.4). Other interviewees who had lost close family

Figure 9.4 The rebuilt village of Ban Nam Khem, on the site of the original village and in a similarly exposed position. Few villagers were living in these houses when we undertook the research in July 2005; instead they continued to sleep each night in temporary housing in the grounds of the local Buddhist monastery.

members were intent on leaving the area entirely, like Pranee from Ban Bang Niang in Khao Lak, who had lost her daughter and who wished to distance herself from her previous life.

Not everyone, of course, was able to make a choice in terms of either their livelihood or their residence and many people in the areas where we worked had to return to fishing and had to live where they had done so previously. Sompong of Ao Hua Laem on Koh Lanta continued to work as a fisherman despite his fear of the sea and of another tsunami. He told us that he would have liked to have moved away from the seashore where he was living because his daughters dared not sleep in the house anymore. But moving house would have made fishing more difficult and he would worry about the security of his boat.

Spatialities of disasters: where does the impact of a disaster begin and end?

Natural disasters usually have sharply defined physical footprints. Whether earthquakes, volcanic eruptions, cyclones, landslides or tsunamis, it is possible to map the areas affected. This is then sometimes coupled with the human context to construct a 'hazardscape' or 'riskscape'. As Cutter,

Mitchell and Scott explain, 'physical hazard exposure and social susceptibility to hazards must be understood within a geographic framework, that is, the hazardousness of a specific place' (2000: 731, and see Cutter, 1996; and Cutter, Boruff and Shirley, 2003). In the 'hazards-of-place' model, it is the combination of the social and biophysical components of vulnerability which produce 'place vulnerability', providing a quite nuanced and fine-grained interpretation of the range of social and economic factors that are at work in producing vulnerability (including, for example, gender, race and ethnicity, employment status, family structure, and social dependence) (Cutter, Boruff and Shirley, 2003: 246-249). In the case of the tsunami, however, almost from the outset we were drawn into seeing the event being connected to people in distant places. Most obviously, the international tourists who were injured or lost their lives caused the tsunami to reverberate beyond the immediate locale, beyond Thailand, and beyond Asia. While this was not a focus of our research, a number of studies have explored the ramifications of the tsunami for survivors and the families of victims in Europe; most of this has focused on post-traumatic stress disorder but we also suggest that, just as in Thailand (see below), livelihoods will have been disturbed as well (see Heir and Meisæth, 2008; Johannesson *et al.* 2009).[9]

What we did explore, however, was the way in which the tsunami, through the evolution of Thailand's space economy, the delocalisation of livelihoods, and the spatial division of households and families, rippled outwards from the littoral space of physical impact to other spaces and social contexts. Traditionally models of risk and vulnerability have assumed a close spatial connection between the physical impacts of cyclones, tsunamis, earthquakes and eruptions on the one hand, and their social and economic impacts on the other, and that this connection is most pronounced in the poorer world. Vulnerabilities to physical hazards, economic risk and social vulnerability have traditionally been understood as relatively closely intersected (see Cutter, Mitchell and Scott 2000; Cutter, 1996; and Cutter, Boruff and Shirley, 2003). Our research demonstrates that this was not the case in southern Thailand where there were not only large numbers of international and domestic tourists, but also very significant numbers of migrants from other regions of Thailand, as well as registered and unregistered migrant labourers from Burma.

Kai was a case in point. Kai, a young tourist worker on Koh Phi Phi, supported a family living in Nakhon Si Thammarat a town some 160 km

---

[9] Interestingly and counter-intuitively, Heir and Meisæth found that having a relative or close friend injured in the tsunami "seemed protective against subsequent health problems" (2008: 274). They speculate that this might be because it protects by 'shielding' or 'distraction'; possibly reduces the feeling of victimhood; or perhaps is linked to greater resilience associated with the demands of assuming a caretaking role.

east of this resort island. Kai was on Phi Phi when the tsunami hit, but managed to escape the wave's force and spent two nights hunkered down in the hills before returning to his home and parents in Nakhon Si Thammarat, traumatised by the experience and what he had seen. There he stayed for three months until economic pressures and the cajoling of his parents encouraged him to return to Phi Phi to find work. At the time we interviewed him in July 2005 Kai was working helping to clear debris from Phi Phi Don's beaches and from the seabed. The dependence of southern Thailand's economy on workers from other regions and countries transformed the tsunami's impact from nodal and focused (as might have been the case had the tsunami struck in 1974 rather than 2004), to networked and dispersed. Not only was southern Thailand dependent on non-local labour, but the natal households of this labour were dependent on income generated by work in places like Khao Lak and Koh Phi Phi. Across Thailand and Burma, sometimes hundreds even thousands of kilometres away from the physical site of the disaster, there were households struggling to recover from the tsunami. In light of this, in contrast to Cutter, Mitchell and Scott's 'hazardscape' (2000) we have suggested the livelihood effects of the tsunami should be conceptualised in terms of a hazard 'footprint':

> Spatially brought together by the diverse forces that shape the modern world, those same forces led to a reverberation of effects back to the homesteads of poor rural families whose migrant sons, daughters, husbands and wives had been killed or injured, and to the apartments and houses of rich cosmopolitans (Rigg *et al.* 2008: 145).

While it is usual to view 'communities as the totality of social system interactions within a defined geographic space such as a neighborhood, census tract, city, or county' (Cutter *et al.* 2008: 599), the community of tsunami victims could not easily be captured in such a geographically defined manner. The networked and spatially dispersed nature of living in Thailand – such that the household is no longer based on the propinquity of its members – meant we could not read off the effects of the tsunami from the site of physical impact alone. This also highlights a difference between the natural science approach, where the littoral context set the boundaries of study, and our research where the information we gleaned took us to other places.

Conclusions

> Society needs to do a better job of asking what kind of tomorrow we create with the possibilities that science offers. Such decisions are governed by values, beliefs, feelings; science has no special voice in such democratic debates about values. But

science does serve a crucial function in painting the landscape of facts and uncertainties against which such societal debates take place (Lord May of Oxford, President's Anniversary Day Address, Royal Society, 2001)

National and global funding agencies, like the National Science Foundation (NSF) who funded the research on which this chapter draws, are slowly recognising the need to fund teams of scholars and researchers that can tackle the complex problems associated with society-environment relations from a variety of disciplinary perspectives. Indeed, the NSF has funded such teams since the inception of its Human and Social Dynamics program in 2004, which supports quick response field research in natural hazards (Suter, Birkland and Larter, 2009). Yet as we reflect on our own experience, we can appreciate the kinds of challenges collaborative work on hazards entails. Various scholars have usefully tried to summarise the problems and possibilities of working in diverse research teams and the differences between cross-, inter-, multi- and trans-disciplinary approaches to research (Table 9.5). These distinctions help specify the different levels of autonomy and dependence that partner disciplines have in the research process/project, and the likely outcomes of such research in terms of whether they are located within disciplinary traditions, or can transcend them.

According to these definitions our own team collaborated in a multidisciplinary way, largely working towards understanding a 'problem' of interest to us as individual researchers, but also to the research funders (the 'impacts' of the tsunami). In this way, the physical scientists contributed to ongoing debates about the physical character and history of tsunamis in the region, while the social scientists spoke to debates about local and global politics, livelihoods, culture and social capital. While there were some attempts to share methodology, this was restricted to a small number of rather peripheral methods: collecting timeframes, eyewitness accounts and photographs. Our interpretation of the materials collected via these methods varied, however, largely because the team worked in different epistemological traditions, with the physical scientists being more positivist/general and the social scientists more critical/contextual (McNeill, Garcia-Godos and Gjerdaker, 2001: 23). At one stage in the fieldwork we did stop to contemplate what broader philosophical questions might connect us, and we began to consider the significance of 'time' to understanding the short and long-term impacts of tsunamis – albeit in different ways (from geological and ecological to human and ephemeral). We did not pursue this, however, as it was not central to the 'problem' set out to be researched or to the funder's project 'deliverables'. When the research was completed, and the funding report written, we largely retreated to our own disciplines to publish our findings, as summarised in

Table 9.5 What's in a name? Disciplinary boundary crossings

Boundary crossing		Characteristics
Disciplinary	Increasingly holistic approach and integration across disciplines →	Research that is undertaken within the intellectual and methodological boundaries of a single discipline
Multi-disciplinary		Research undertaken when disciplines work together on a problem or real world issue but maintain their autonomy in terms of methods, conceptual framings and theoretical structures. Such an approach juxtaposes rather than combines disciplines, and therefore does not transcend disciplinary boundaries.
Cross-disciplinary		Research based on a common theoretical understanding and accompanied by a mutual interpenetration of disciplinary epistemologies
Inter-disciplinary		Research which involves cooperation and the combination or integration of ideas, data, techniques, concepts and methods across at least two disciplines so as to develop a common approach and framework of research. Klein (2007: 37-8) distinguishes between narrow inter-disciplinarity in which two allied disciplines work in tandem, such as social anthropology and human geography, and wide inter-disciplinarity when disciplines in very different fields integrate, such as social anthropology and earth sciences.
Trans-disciplinary		Research that 'as a practice' (Cronin 2008: 3) attempts to challenge and transcend disciplinary boundaries. It not only integrates ideas across disciplines but in so doing generates new ideas, data, methods and approaches. In addition to integrating across subject areas, trans-discplinary research also brings together academic and non-academic partners, working in the public, private and civil society sectors.

Adapted from: McNeill, Garcia-Godos and Gjerdaker, 2001; Cronin, 2008; and Klein, 2007

Table 9.4.[10] This was at least partly because research assessment exercises (such as those in the UK and Australia) tend to privilege disciplinary journals – constituting yet another barrier to building truly inter-disciplinary research traditions and practices.

In geography, perhaps the cross-disciplinary subject *par excellence,* there are long-standing debates about how best to foster greater collaboration and mutual working – or inter-disciplinary research – and what has impeded such work. Thrift (2001) has identified a respect and trust deficit

[10] It is also notable how the natural and social scientists ascribed authorship. In the science tradition, the natural scientists tended to include all those who were members of the research project whether or not they had had a direct involvement in the paper (Horton *et al.* 2008); the social scientists, meanwhile, only included those who had contributed to the writing of the paper (e.g. Rigg *et al.* 2008 and Tan Mullins *et al.* 2007).

between physical and human geographers; Jones and Macdonald (2007) an 'epistemological tension'; Cook and Lane note how difficult it is to build bridges across different communities of practice; while Bracken and Oughton (2007) focus on issues of language and the challenge of talking across intra-disciplinary boundaries. With regard to their research on risk and flood management in Bangladesh, Cook and Lane conclude by writing:

> Despite shared faith in interdisciplinarity and growing recognition for the complexity of environmental problems, there is very little consideration for the practical difficulties of integrating disparate knowledge claims.... There is no discussion of the practical realities of consolidating knowledges from different fields or for the difficulties that will accompany the communication of such ideas to managers or to the public. Implicitly, these claims assume that 'all the relevant' knowledges can be collected, reconciled and understood as part of a 'holistic managerial collage' (Cook and Lane, 2010: 21).

The challenge of building truly trans-disciplinary risk research, even inter-disciplinary risk research, applies not just to the academy. It is equally germane in the spheres of risk management and the governance of risk where many of the same barriers touched on in this chapter are repeated.

Acknowledgements

The authors gratefully acknowledge the financial support of the US National Science Foundation (NSF project number 0522133). In addition, we would like to express our thanks to the late Sophia Buranakul and to Wendy Firlotte in Krabi, the Prince of Songkhla University, to all our respondents and to the staff of government and nongovernment agencies and organisations who gave their time to talk to us and who so willingly spoke about their often traumatic and sometimes tragic experiences. This chapter was completed while Jonathan Rigg was a Gledden Senior Visiting Fellow at the University of Western Australia in Perth and he would like to acknowledge the support of the Institute of Advanced Studies at UWA, Dr Brian Shaw and the School of Earth and Environment.

References

Anderson, N. B. (1999) Solving the puzzle of socioeconomic status and health: the need for integrated, multilevel, interdisciplinary research. *Annals of the New York Academy of Sciences*, **896**, 302–312.

Bracken, L. and Oughton, E. (2006) 'What do you mean?' The importance of language in developing interdisciplinary research. *Transactions of the Institute of British Geographers*, **31**, 371–382.

Brewer, G.D. (1999) The challenges of interdisciplinarity. *Policy Sciences*, **32**(4), 327–337.

Brown, P. and Minty, J. (2006) Media coverage and charitable giving after the 2004 tsunami. William Davidson Institute Working Paper Number 855, December. The William Davidson Institute at the University of Michigan. Downloaded from: http://wdi.umich.edu/files/publications/workingpapers/wp855.pdf

Buranakul, S., Grundy-Warr, K., Horton, B., Law, L., Rigg, J. and Tan-Mullins, M. (2005) The Asian tsunami, academics and academic research. *Singapore Journal of Tropical Geography*, **26**(2), 244–248.

Cook, B.R. and Lane, S.N. (2010) Communities of knowledge: Science and flood management in Bangladesh. *Environmental Hazards*, **9**, 8–25.

Cronin, K. (2008) Transdisciplinary research (TDR) and sustainability. Overview report prepared for the Ministry of Research, Science and Technology (MoRST), New Zealand. Downloaded from: http://learningforsustainability.net/pubs/Transdisciplinary_Research_and_Sustainability.pdf.

Cutter S.L., Mitchell, J.T. and Scott, M.S. (2000) Revealing the vulnerability of people and places: a case study of Georgetown County, South Carolina. *Annals of the Association of American Geographers*, **90**(4), 713–737.

Cutter, S.L. (1996) Vulnerability to environmental hazards. *Progress in Human Geography*, **20**(4), 529–539.

Cutter, S.L., Boruff, B.J, and Shirley, W.L. (2003) Social vulnerability to environmental hazards. *Social Science Quarterly*, **84**(2), 242–261.

Cutter, S.L., L. Barnes, M. Berry, C. Burton, E. Evans, E. Tate and J. Webb (2008) A place-based model for understanding community resilience to natural disasters. *Global Environmental Change*, **18**(4), 598–606.

FHI (nd) Qualitative Research Methods: A Data Collector's Field Guide. Family Health International. Downloaded from: http://www.fhi.org/nr/rdonlyres/etl7vogszehu5s4stpzb3tyqlpp7rojv4waq37elpbyei3tgmc4ty6dunbccfzxtaj2rvbaubzmz4f/overview1.pdf.

Gieryn, T.F. (1983) Boundary-work and the demarcation of science from non-science: strains and interests in professional ideologies of scientists. *American Sociological Review*, **48**(6), 781-795.

Hakoda, T. (2005) Invisible victims of the tsunami – Burmese migrants workers in Thailand. *FOCUS*, **39**(March). Downloaded from http://www.hurights.or.jp/archives/focus/section2/2005/03/invisible-victims-of-the-tsunami–burmese-migrant-workers-in-thailand.html.

Heir, T. and Meisæth, L. (2008) Acute disaster exposure and mental health complaints of Norwegian tsunami survivors six months post disaster. *Psychiatry*, **71**(3), 266–276.

Johannesson, K. B., Michel, P.O., Hultman, C.M., Lindam, A., Arnberg, F. and Lundin, T. (2009) Impact of exposure to trauma on posttraumatic stress disorder symptomatology in Swedish tourist tsunami survivors. *Journal of Nervous & Mental Disease*, **197**(5), 316–323.

Jones, P. and Macdonald, N. (2007) Getting it wrong first time: building an interdisciplinary research relationship. *Area*, **39**, 490–498.

Klein, J.T. (2007) Interdisciplinary approaches in social science research, in *The Sage Handbook of Social Science Methodology* (eds W. Outwaite and S.P. Turner), Sage Publications, Los Angeles, pp 32–49.

Lebel, L., Supaporn Khrutmuang and Manuta, J. (2006) Tales from the margins: small fishers in post-tsunami Thailand. *Disaster Prevention and Management*, **15**(1), 124–134.

McNeill, D., Garcia-Godos, J. and Gjerdaker, A. (2001) Interdisciplinary research on development and the environment. SUM Report No. 10, Centre for Development and the Environment, University of Oslo. Downloaded from: http://www.sum.uio.no/pdf/publications/reports/sum-report-serie/sum-report_10.pdf.

Merli, C. (2010) Context-bound Islamic theodicies: The tsunami as supernatural retribution vs. natural catastrophe in Southern Thailand. *Religion*, **40**, 104–111.

Mills, M.B. (1995) Attack of the widow ghosts: gender, death, and modernity in Northeast Thailand. In *Bewitching Women, Pious Men: Gender and Body Politics in Southeast Asia* (eds A. Ong and m.G. Peletz), University of California Press, Berkeley, pp. 244–273.

Pattana Kitiarsa (2005) Beyond syncretism: hybridization of popular religion in contemporary Thailand. *Journal of Southeast Asian Studies*, **36**, 461–487.

Piker, S. (1968) The relationship of belief systems to behavior in rural Thai society. *Asian Survey*, **8**(5), 384-399.

Rigg, J., Law, L., Tan-Mullins, M. and Grundy-Warr, C. (2005) The Indian Ocean tsunami: socio-economic impacts in Thailand. *The Geographical Journal*, **171**(4), 374–379.

Rigg, J., Tan-Mullins, M., Law, L. and Grundy-Warr, C. (2008) Grounding a natural disaster: Thailand and the 2004 tsunami. *Asia Pacific Viewpoint*, **49**(2), 137–154.

Suter, L., Birkland, T. and Larter, R. (2009) Disaster research and social network analysis: Examples of the scientific understanding of human dynamics at the National Science Foundation. *Population Research Policy Review*, **28**, 1–10.

Tan-Mullins, M. Rigg, J. and Grundy-Warr, C. (2008) Responses and resilience of fisherfolks on the tsunami event in Southern Thailand, in *Crucible for Survivial: Environmental Security and Justice in the Indian Ocean Region* (eds T. Doyle and M. Risely), Rutgers University Press, New Jersey, New York and London, pp. 116–129.

Tan-Mullins, M., Rigg, J., Law, L. and Grundy-Warr, C. (2007) Mapping the local politics of aid: structures, networks and capitals in post-tsunami Thailand. *Progress in Development Studies*, **7**(4), 327–344.

Thomas, A. and Fritz, L. (2006) Disaster relief Inc. *Harvard Business Review*, November, 114–122. Downloaded from: http://errrmsystems.pbworks.com/f/thomasandfritz.pdf.

Thrift, N. (2002) The future of geography. *Geoforum*, **33**, 291–298.

Tsuji, Y., Sumatra International Tsunami Survey Team (2005) The 26 December 2004 Indian Ocean Tsunami: Initial Findings from Sumatra. Tsunamis and Earthquakes, Western Coastal and Marine Geology, USGS. http://walrus.wr.usgs.gov/tsunami/sumatra05/heights.html.

Yang, B.Y.F. (2007) Life and death away from the Golden Land: the plight of Burmese migrant workers in Thailand. *Asian-Pacific Law and Policy Journal*, **8**(2), 486–535.

CHAPTER 10

Social Work in Times of Disaster: Practising Across Borders

Lena Dominelli
University of Durham, Durham, UK

Introduction

Social workers have important roles to play in disaster interventions, providing humanitarian aid on the ground, both in their own countries and overseas. Social workers were given this task because they, of all the 'social' professions that serve people, purport to 'enhance the well-being of individuals, groups and communities' (www.iassw-aiets.org) as a normal part of their daily work. Building human and infrastructural capacity that has been damaged by catastrophic events also enters their remit. Practitioners claim to work according to egalitarian, empowering values that promote human well-being through the realisation of a human rights and social justice based practice (Ife, 2001; Dominelli, 2002). A professional risk is the gap between professionals' espoused goals and their actual delivery. This is not unexpected given that social workers do not directly control people's behaviour 24 hours a day. Rather, they assess and negotiate risks and operate within the boundaries of persuasion and legislative frameworks that restrict their autonomy and authority to act within tightly defined legal procedures and professionally agreed ethical codes.

Social workers' voices are often missing from public discourses before, during and after catastrophe strikes. This silence contrasts with the media spotlight when interventions go wrong, as they do in child protection cases where carers murder a child as occurred to British toddler Peter Connelly, killed by his mother, her boyfriend and boyfriend's brother in August 2007 (Laming, 2009). Social workers 'failure to protect' a child signifies a failed risk assessment that endangered a child's life. Although their inability to safeguard children's lives is a matter of serious concern, the inadequacy of the tools for making risk assessments in complex situations involving a number of actors was not raised in the media furore

Critical Risk Research: Practices, Politics and Ethics, First Edition.
Edited by Matthew Kearnes, Francisco Klauser and Stuart Lane.

that ensued. The media focused exclusively on holding social workers accountable for their actions. Interestingly, a similar outcry did not occur for the police or health professionals who held similar responsibilities in this case. Public anger targeting social workers is not unusual in Britain, having been raised in more than 80 child protection inquiries since the first one in 1973 when Maria Colwell was killed by her stepfather in Brighton (Parton, 2004). But the levels of hysteria reached by *The Sun* over Peter Connelly's death were unprecedented.

Social workers are involved throughout disaster relief processes – assessing need, co-ordinating and delivering goods and services, facilitating family reunification, counselling those bereaved in the immediate aftermath and supporting individuals and communities in rebuilding their lives, developing resilience and building capacity to minimise future risks. Risk minimisation is risky because disaster interventions depend on the nature of the disaster, local conditions and traditions, and personnel and resources available for particular activities. The presence of many unknown factors in practice interventions makes risk assessment and management in social work an uncertain science (Quinsey, 1995; Dominelli, 2010). However, it may be possible for practitioners to develop more robust risk assessment instruments by collaborating with risk researchers in the physical sciences. Current disaster practices tend to ignore issues of social justice and risk assessment tools devised to assist practitioners in their tasks individualise social problems, blame individuals for having them, and neglect links between personal behaviours and social structures and individual positions within them (Swift and Callahan, 2010). The profession can enhance its credentials in this arena by liaising with physical scientists holding risk research expertise. The latter might become embroiled in the 'politics of research' when theory and practice interact in unanticipated directions (Dominelli, 1998).

At the same time, social workers working within an ethical, human rights and justice based framework for practice can share their insights with physical scientists. Included in this is the notion of differential interventions to achieve equitable outcomes. Disasters affect countries differently and each nation and group within it has variable capacities or resilience in responding to the problems created. Low income countries and poor people have greater difficulties addressing disasters. They find it hard to adjust to disasters including climate change without having contributed much to creating them (UNDP, 2007, 2008). Disasters impact mainly on women and children (UNDP, 2008) who are least likely to have a direct say in interventions affecting them (Pittaway *et al.*, 2007).

In this chapter, I consider social workers' roles in disasters at home and abroad, identifying the principles of empowering practice that support local residents and engage them in resilience building, community development and decision-making processes before, during and after disasters. I argue that social workers' practice in such situations could improve through multidisciplinary partnerships that involve dialogues across many borders, especially those erected by disciplinary and professional boundaries. Social workers and physical scientists including engineers, geologists and seismologists should dialogue across disciplinary controversies. Their separate knowledges could have a greater impact if they talk to each other, learn from each other and improve their practices through such exchanges.

I focus on social workers' roles in disasters as professionals working with risky uncertainties from their specific disciplinary base. I critique such practices to identify strengths and weaknesses and consider how multidisciplinary approaches in risk research could improve practice during the prevention, immediate recovery and long-term reconstruction phases of disaster relief. These build on knowledge gained through the Rebuilding People's Lives Network (RIPL) organised by the International Association of Schools of Social Work and when supporting survivors of the 2004 Indian Ocean Tsunami; earthquake survivors in Sichuan (2008), Haiti (2010), Chile (2010), Christchurch (2010) and Japan (2011); and interdisciplinary approaches demonstrated through the projects associated with these disasters.

Social workers' roles during times of disasters

Defining disasters

Disasters are natural and human-made phenomena occurring as unpredictable events that have horrendous outcomes including the destruction of lives, property and physical environment. Their impact may be localised, or wide-ranging. Even if limited in scope, boundaries between the local and the global can become blurred. The chemical disaster in Bhopal, India during the 1980s was a local event with international repercussions because a global company was involved and individuals' claims for compensation reached the American courts. Recently, the failure of nuclear reactors in Fukushima Daiichi following Japan's 2011 earthquake tsunami caused several western nations to rethink their proposed expansion in nuclear energy. For example, Angela Merkel suspended an extension of the lifetime of Germany's existing nuclear plants.

Perez and Thompson (1994) have defined disaster as 'widespread extensive damage that is beyond the coping capacity of any community and therefore requires external intervention'. The United Nations (UN) and the World Health Organisation (WHO) use Gunn's (1990) definition of disaster as 'the result of a vast ecological breakdown in the relationship between man [sic] and his [sic] environment... [with] disruption on such a scale that the stricken community needs extraordinary efforts to cope'. Disaster interventions have traditionally focused on flooding, tsunamis, landslides, hurricanes, earthquakes, volcanic eruptions and drought. These are associated with nature and described as 'natural disasters'. Those linked to industrial pollution or environmental degradation, e.g., Bhopal, and armed conflicts like that which occurred in Rwanda, or mass murders similar to those inflicted upon Columbine High School Massacre of 1999 in the USA are termed '(hu)man-made'. Social workers intervene in both types, to provide food, clothing, access to medical aid, shelter, reunification of families, counselling support and community development initiatives to rebuild communities.

Dominelli (2007) has suggested that social workers enlarge existing definitions of disasters to cover poverty because it is the largest human-made disaster; and climate change because it differs from other ecological disasters (Dominelli, 2009). Poverty causes disasters and exacerbates its effects. It undermines people's resilience and capacity to cope individually and collectively, e.g., in 2005 when Hurricane Katrina hit the Lower Ninth Ward in New Orleans in the USA. Occurring in the richest country in the world, it disproportionately affected poor African American communities who bore the brunt of the $140 billion of damages and were least able to deal with its effects in both short and long-terms (Pyle, 2007). Those most seriously affected by disasters whether natural or human-made, include indigenous people and those on low incomes. The UN General Assembly approved the Declaration on the Rights of Indigenous Peoples in 2007 to address such issues.

Climate change will add to the numbers affected by disasters if temperature increases are not kept to below 2°C (Stern, 2006) and disproportionately affect poor people, especially those leading nomadic lifestyles in sub-Saharan Africa or living in low-lying areas like Bangladesh. The UN estimates that an additional 250 million people might have to move by 2050 if temperature rises are not reduced (Loescher *et al.*, 2008). Climate induced disasters are caused by extreme weather events linked to burning fossil fuels and industrialisation (Dessler and Parsons, 2010). The effects of climate change include: the disappearance of small island nations in the Pacific Ocean like Tuvalu and the Maldives because increased temperature will melt glaciers and ice-caps in the Arctic and Antarctic and raise ocean levels significantly; drought and the desertification of land, especially in Sub-Saharan Africa; food insecurities for half a billion more people; and water shortages for 1.4 billion in India and China affected by glaciers melting in Nepal. Such events will intensify mass migrations within and between countries (UNDP, 2007, 2008, Sanders 2009). Social workers are responsible for working with people fleeing climate-change based disasters and armed conflicts over natural resources locally, nationally and internationally (Ramon, 2008).

Defining risk and resilience

During their interventions, practitioners draw upon two key concepts: risk and resilience, to reduce risk and enhance resilience amongst individuals and communities. Both concepts are firmly rooted in the physical sciences and have been imported into social work without much dialogue between these disciplines to ensure that both groups could transcend disciplinary protectionism and benefit from each others' insights. Fortunately, the beginnings of mutual discourses are occurring in universities like Durham which has the Institute of Hazard, Risk and Resilience Research (IHRR) engaging in cross-disciplinary disaster research.

Risk in social work is 'part of everyday life . . . most closely associated with fear and the efforts that people make to increase their security and safety' (Swift and Callahan, 2010, p. 9). Risk measurement and assessment tools have been used extensively in controlling criminal behaviour, child protection cases, and mental health work where protecting people from harm is central to risk management concerns. Here, practitioners aim to eliminate or reduce potential dangerousness in the behaviours of perpetrators and safeguard lives and property. Their emphasis, therefore, is on behaviour and not physical or environmental structures. Social workers deemed the calculation of physical risk in disaster interventions the prerogative of the physical sciences and engineers who had the expertise to assess the probabilities of earthquakes, landslides or volcanic action occurring and deferred to their expertise. In taking for granted that others were addressing the infrastructural implications of risk research for people's lives, social workers missed the opportunity to become informed about physical risks and their relevance for and potential impact upon practice. Thus, there is a gap in their knowledge that requires attention. Similarly, one might add, there is information about human behaviour that physical scientists need. Consequently, dialogue and risk research projects involving them both can be mutually beneficial.

The situation began to shift in 2009 when social work began to explore potential cross-disciplinary links with physical scientists during the climate change talks in Copenhagen, where social workers devoted a whole day's seminar to exploring potential links and led to the formulation of a climate change and disaster intervention policy that received the support of social work's key international organisations, the International Association of Schools of Social Work (IASSW), International Council for Social Welfare (ICSW) and International Federation of Social Workers (IFSW). It also produced an enthusiasm for research involving physical and social scientists, a goal that is currently being facilitated by a number of individuals and groups. For example, in the UK, the United Kingdom Research Council (RCUK) is supporting interdisciplinary research on natural hazards under its Global Uncertainties Programme in a cross-research council initiative involving the Natural Environmental Research Council (NERC) and the Economic and Social Sciences Research Council (ESRC) in funding scoping studies on earthquakes and volcanic activity through the Increasing Resilience to Natural Hazards (IRNH) programme. The Engineering and Physical Sciences Research Council (EPSRC) has an initiative on climate change that includes a project that examines health and social care provisions for older people under conditions of climate change. These ventures are innovative in that they involve collaborative partnerships between physical scientists, engineers, geographers and social scientists including social workers where knowledge and expertise is shared in a mutually supportive environment. These could provide models for working across disciplinary divides and maximising the impact of risk research.

Resilience, the other concept familiar to social workers, is a contested, eclectic and elastic term. Resilience has shifted from its accepted meaning in the physical sciences as the capacity of materials to respond to stress, to the social sciences, arts and humanities where this definition is used uncritically to manage crises (Manyena, 2006). Used as a systems management tool, resilience is in danger of losing its potential to enhance sustainability and create a step-change in developing resilient resources that enable people, communities, institutions and organisms to thrive in contexts of unremitting and/or threatening change.

Through critiques of this usage, resilience has re-emerged as an active concept referring to the capacity of systems, whether natural, human or hybrid, to sustain themselves in the face of endogenous and exogenous shocks to an existing state. Resilience expressed in this way has a wide-ranging and general appeal as it rebalances and refocuses the primacy traditionally given to managing system shocks. Also, resilience is non-linear in that a system might be resilient along one dimension, but not

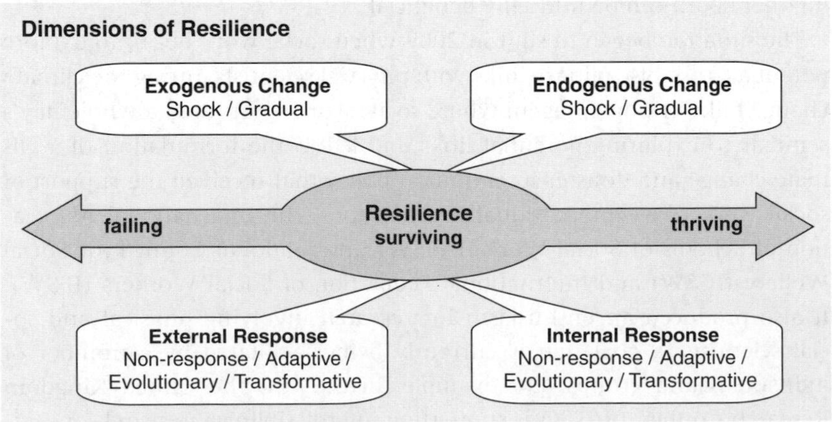

Figure 10.1 Dimensions of Resilience Chart

in another. Social workers suggest that this is crucial in systems involving people who act as agents and evaluate many dimensions of risk, often intuitively, before choosing a particular course of action. Additionally, resilience responses operate on a continuum that ranges from non-responses, which in some cases might turn out to be resilient, to survival through adaptation, accommodation or making do with a situation, to thriving responses that are transformative. These can produce structural changes to systems, innovations in thinking about situations and developing something new and different. I have tried to articulate this nuanced complexity in the chart above.

The history of social work in disaster interventions: strengths and limitations of expert-led, top-down approaches

Disaster intervention manuals have traditionally focused on reducing risk defined as minimising the objective and subjective probability that a negative event will occur. These suggest that the following actions are crucial in and provide a basis for social workers' interventions in disaster relief situations:

- Reversing any ill affects on health.
- Modifying identified hazards.
- Decreasing vulnerabilities and increasing resilience.
- Improving disaster preparedness for future possible disasters (Perez and Thompson, 1994).
- Empowering local residents by responding to their needs as they define them in a locality specific, culturally relevant manner (Dominelli, 2010).

Social workers initially became involved in providing succour after disasters as part of the Marshall Plan for rebuilding Europe after the devastation of the Second World War, when the United Nations High Commissioner for Refugees (UNHCR) was created in Geneva on 14 December 1950. It assumed many functions of the UN Relief and Rehabilitation Administration (UNRRA) including the roles of co-ordinating, overseeing and monitoring developments. It recognised human rights as articulated through the Office of the High Commission on Human Rights (UNHCHR) which was replaced in 2006 by the UN Human Rights Council (UNHRC).

The UNHRC was created as a subsidiary body of the UN General Assembly to address criticisms of the UNHCHR articulated primarily by the USA and Israel for allowing nations with poor human rights records to sit on its decision-making structures. They continue with these criticisms, having voted against its formation alongside two other small nations. George W Bush also boycotted UNHRC deliberations – an irony given the controversies he courted over human rights abuses at Guantanamo Bay in Cuba and under the provisions of the Patriots' Act in the USA (Pearlstein and Posner, 2009).

Given the urgency of the responses and enormity of the task in rebuilding war-torn Europe, the models for practice adopted by UNHCR were top-down and expert-led. This model still prevails, despite various critiques that have emerged over the years and called for bottom-up approaches that empower local residents in making decisions about what should happen in their lives and communities. The most trenchant attack upon expert-led interventions was set out by Hancock (1991) in his book, *Lords of Poverty*, which calls humanitarian aid a multi-billion dollar business that makes profits for a few donors while disempowering recipients.

The UN operates in disaster interventions through various bodies that it established, e.g., the UNHCR and civil society organisations like the International Federation of the Red Cross and Red Crescent Societies (IFRC) which are particularly active in delivering aid in the immediate aftermath of disasters. These focus on providing food, water, shelter and medical supplies on a non-discriminatory basis. IFRC is composed of 186 societies at national level and has 97 million volunteers world-wide. It has been in existence in some form since 1919, although the current structure was adopted in 1991. The Seville Agreement of 1997 was formulated to lessen tensions between the International Committee of the Red Cross (ICRC) and Red Cross and Red Crescent Societies at country level.

The UNCHR, like its predecessor, relies on the IFRC and other NGOs to deliver humanitarian aid. The UNCHR is responsible for reviewing human rights including the Universal Periodic Review, and assessing human rights observances in all 192 member states of the UN. The performance of the UNCHR's work is facilitated by an Advisory Committee and a Complaints Procedure. The Advisory Committee is composed of 26 elected experts in the human rights arena. The Complaints Procedure is conducted through a panel of five representatives, each one being drawn from one of the UN's five regions. Special rapporteurs also undertake country-based investigations into human rights violations. In 2006, the special rapporteurs addressed the issue of poverty during Human Rights Day, thereby affirming the link between poverty and lack of resilience in disaster situations.

The UNHCR supports the work of the UNCHR by being responsible for the 1951 Geneva Convention on refugees that provides refuge for those seeking to escape persecution. The UNHCR is responsible for 36.4 million people; most are internally displaced. Only 800 000 of the overall total are asylum seekers. None of those included in the UNHCR figures are people seeking refuge from climate change because they are not covered by the 1951 Geneva Convention (Loescher *et al.*, 2008). The UNHCR has expanded humanitarian aid to cover 'persons of concern' as defined by the 1951 UN Convention Relating to the State of Refugees which includes internally displaced persons, the 1967 Protocol and the 1969 Organisation for African Unity Convention.

Other UN bodies associated with disaster interventions are the UN Disaster Assistance and Co-ordination (UNDAC) covering 57 countries that include the Philippines, Vietnam, Cambodia, Samoa, and Tonga. Additionally, international NGOs or parts of civil society like the IFRC, Oxfam and Save the Children are active in it. In the immediate aftermath of disasters, organisations like these provide urgently needed food, clothing, shelter and medicine, using practitioners and volunteers to distribute these amongst disaster survivors. In 2008, the IFRC responded to 326 natural disasters that killed 235,736 people, the highest level since the 2004 Indian Ocean Tsunami (IFRC, 2009). These interventions depict social work in action, but many call themselves relief workers rather than social workers, raising questions about the appropriation of practice by those lacking social work qualifications and training. Collaboration with physical scientists including those doing risk research might intensify such concerns. These differentials also indicate status differences between disciplines in the academy where the physical sciences act as the benchmark for the others. Social work research has never received the significant sums allocated to its endeavours that others such as health enjoy in either the research councils or government departments. Social work academics challenged this state of affairs by developing a 20 year research strategy in 2008 and argued for substantial funding to redress this historical anomaly (Bywaters, 2008). Action to demand specific recognition of social work as a research-led discipline was initiated in 1996 and realised for social work training in 2005 and for research projects in 2007 (Dominelli, 2004, 2009).

The current UN Under-Secretary for Humanitarian Affairs is Valerie Amos who replaced John Holmes in 2010. She coordinates the Office for the Coordination of Humanitarian Affairs (OCHA) which replaced the Department for Humanitarian Affairs in 1998. OCHA has an Executive Committee for Humanitarian Affairs, an Emergency Relief Coordinator and the Inter-Agency Standing Committee (IASC). The UNHCR has staff in 110 countries to implement its responsibilities. The original remit for IASC arose through the General Assembly in 1992 when Resolution 46/92 was passed. This developed the infrastructure to achieve better: coordination of activities and resources; inter-agency decision-making in complex emergencies; and integrated multi-sectoral approaches to disaster relief. IASC is composed of: UN agencies like the UNDP, UNFPA, UNHABITAT, UNHCR, UNICEF, WHO and World Bank; standing invitees like ICRC, IFRC; and other civil society organisations known for their work in humanitarian aid that are invited on an ad hoc basis. The IASC has a Working Group to develop guidelines for interventions during disasters. These aim to improve coordination and facilitate inter-agency decision-making while providing guidelines that uphold human rights, ethical behaviour and empowering values. These cover a range of subjects including community development, women and mental health. Social workers led in the creation of those linked to psycho-social interventions (Bragin, 2008). OCHA is also linked to the Consolidated Appeals Process (CAP) and the Central Revolving Emergency Fund (CERF).

Other forms of UN support for humanitarian aid during different disasters include peacekeeping activities in 24 countries including Afghanistan, Iraq, the Democratic Republic of the Congo, Darfur, Somalia and Kenya. The UN High Commission for Refugees' Representation in Cyprus is charged with assisting residents of that troubled island. The UN Relief and Works Agency for Palestine Refugees in the Middle East (UNRWA) has been responsible for implementing the UN's Palestine Mandate and caring for refugees resulting from the formation of Israel. The conflict between Israelis and Palestinians is ongoing, with limited chances of reaching a solution over human rights violations and land claims. Palestinians who are not covered by UNRWA can access UN support through the UNHCR's more limited provisions for covering refugee status. The UN's Food Programme is another initiative that assists in disasters by providing food aid. It currently provides food to 100 million people in 80 countries.

IASSW, formed in 1928 by educators, was responsible for training and educating practitioners involved in these activities. IASSW has had consultative status with the UN since 1947 and played an important role in discussions around the Universal Declaration of Human Rights (UDHR). Articles 22 to 27 are closely linked to individuals' rights to food, clothing, shelter, health, education and social services provisions and can be used by social workers in arguing for resources when responding to human need in disaster situations because all countries that are members of the UN are signatories. IASSW is also committed to observing UN Conventions and Covenants. IASSW works with the International Federation of Social Workers (IFSW) to strengthen the voice of social work in the international arena, where they are seen as small players in disaster interventions, although they have members in 84 countries. Both work with the International Council for Social Welfare (ICSW), the other 'sister' organisation for the profession.

Social workers' roles in disaster interventions

The UN sees social workers' roles primarily as mobilising communities to: assess situations; identify need, distribute resources in the immediate aftermath of disasters; and provide psycho-social care. The IASC included social workers in developing psycho-social guidelines (Yule, 2008; Bragin, 2009), but social workers' voice at the decision-making table is weak.

Their professional associations encourage social workers to respond to disasters within environmentally friendly, human rights and social justice perspectives. Social workers' roles in disaster situations are discipline specific and usually enacted in the immediate aftermath of a disaster. They include being a:

- Facilitator who gets different groups to talk to each other, including in the co-production of knowledge.
- Coordinator of agencies providing resources and assistance.
- Community mobiliser (of people and systems) in the immediate aftermath of disaster and the longer-term.
- Mobiliser of local resources (people and materials) based on assessments of need, often supplementing these by bringing in external resources.
- Negotiator or broker between communities and different levels of government on behalf of people excluded from discussions about distributing aid and planning future developments.
- Mediator between conflicting interests and groups, including those based on gender relations.
- Consultant to government and other agencies, identifying and representing the needs and interests of affected communities.
- Advocate for the rights and entitlements of marginalised individuals and groups during the distribution of aid and prevent their voices being ignored.
- Educator, giving out information about accessing relief, caring for oneself, and avoiding the spread of diseases following disasters.
- Trainer in mobilising local resources and responding effectively when disaster strikes.
- Cultural interpreter for external relief workers and donor agencies who may not be familiar with local cultural traditions and languages.
- Family reunification worker helping to find lost family members.
- Counsellor to those grieving the loss of family, friends, neighbours, homes and communities.
- Therapist helping people deal with the emotional consequences of disaster (Dominelli, 2009).

Recent disasters indicate that practitioners must do more than follow social work roles and principles to truly empower local residents. They must gain knowledge about the physical terrain in which they work and the construction of disaster-proof buildings and the built infrastructure. Currently, they lack such information, but they could access it through partnerships with the physical sciences. For example, in Sichuan, seismologists Li *et al.* (2010) had information about fault-lines carrying significant implications for the location of buildings following reconstruction. Had social workers known about these potential hazards, they could have used their communication skills to translate these to local people and deter building on fault-lines by pointing out how the physical terrain could undermine safety in future earthquakes in locally accessible language, as is being done by NSET (National Society for Earthquake Technology) and the Nepal School of Social Work in Nepal. In Haiti, practitioners could show that relief camps sited by potential landslides face increased hazards during heavy rains and advise that temporary shelters be erected elsewhere. In Chile, social workers could focus on accessing resources when poverty prevents residents from earthquake-proofing housing as stipulated by law and through such inaction unknowingly exacerbate vulnerability during subsequent earthquakes. The knowledge available through the physical sciences is of little use to those of low income without strategies to overcome the lack of funds. Collaboration with social workers can ensure that existing scientific knowledge is used.

Social workers, as translators of physical science knowledge in community settings, could convey the risks associated with particular courses of action during preventative and reconstruction phases of intervention and thereby add the role of interdisciplinary translator and communicator to those identified above. Lane (2011) describes effective translation and communication work in co-producing community-based solutions to flooding problems in North Yorkshire. Such endeavours could be strengthened through collaborations involving community groups and physical and social scientists including social workers and community development workers devising action plans and conducting risk assessments together.

Building on Desai (2007), I posit the skills and processes for social workers working with disaster-affected communities below. They do not necessarily follow one another in a neat order, but occur simultaneously and messily. They are:

- Making initial contact with victim-survivors.
- Engaging local people and seeking their opinion throughout the processes of intervention.
- Assessing available information and identifying needs and areas of work.
- Drawing up contracts for work to be undertaken.
- Initiating the required actions and monitoring service delivery to ensure these reach those requiring assistance.
- Constantly evaluating the results of interventions.
- Resisting top-down interventions and supporting bottom-up ones devised in partnerships that fully involve local residents.
- Constructing bridges for discourses that cross disciplinary divides, especially between engineers and physical scientists and community groups to engage residents fully in co-producing knowledge and solutions to the problems that disasters create.

The values of fairness and justice are essential underpinnings for disaster relief efforts involving social workers. Finding solutions that implement the objectives identified above within fair, equitable and culturally sensitive responses is very complicated. These require communities and individuals to develop skills in assessing vulnerabilities and risk, adaptive capacities and resilience in specific situations, locations and contexts. Risk researchers can assist practitioners in the development of needs and risk assessment tools to cater for the complex situations encountered in the field. Such developments can also extend to the formulation of preventative measures that affect social care provisions, hospital and other health services and shelter in the short and long-terms. Preventive initiatives can mainstream prevention; early warning systems; and humanitarian responses. These actions can be strengthened by engaging social workers in discussions with physical scientists about how best to reach communities. Such long-term goals may not materialise easily in disaster conditions when basic communication, transportation and governance infrastructures have disappeared. Collaboration between physical and social scientists could assist those on the ground to address these deficiencies more speedily and realise the goals of facilitating ethical practice that is truly participative and empowering – something that policymakers have failed to achieve despite their rhetoric on community engagement (Pattison, 2007; Popple, 2007). Recipients' rights to services under Articles 22-27 of the UDHR are violated when needed services are absent. Social workers can highlight these through empirical research that involves physical scientists to identify the higher risks that people encounter and greater costs incurred by society if it reinforces or exacerbates social exclusion when disaster strikes. For example, people living in areas affected by the 2004 Indian Ocean Tsunami remained in temporary accommodation in 2011, but this was not known beyond their village until researchers exposed their plight. Disaster endeavours have to uphold human rights, social and environmental justice, and the ethical and equitable distribution of finite resources to create robust, empirically-grounded solutions that will survive into the future.

Another difficulty UN humanitarian aid efforts social workers have to transcend arises from two competing and conflicting principles involved in disaster responses: state sovereignty; and the state's duty to protect residents. Article 2(1) of the UN Charter restricts international action in the internal affairs of member states by affirming the Montevideo Convention on the Rights and Duties of States that recognised state sovereignty as fundamental to international relations as stipulated by the League of Nations in 1933. The 'Responsibility to Protect', first promoted by the International Commission on Intervention and State Sovereignty in 2000, can provide an antidote to the 'sovereignty' principle when invoked by some members of the UN to prevent assistance being delivered by external agencies. This occurred, for example, in 2008 in Myanmar/Burma when Cyclone Nargis devastated large swathes of the country. Colonel Gaddafi has argued similar points in Libya.

The UN Security Council (UNSC) can threaten or use 'armed force without the agreement of the target state to address humanitarian disasters caused by grave, large-scale and fundamental human rights violations' (Perez and Thompson, 1994). The UNSC can authorise intervention in humanitarian disasters where serious, widespread damage, considerable loss of life and gross human rights violations are evidenced. However, it is reluctant to impose sanctions against any specific state. At the same time, UN personnel delivering aid run the risk of being abducted, abused and/or killed for embarking on humanitarian work, because they are seen as part of the problem, not the solution by providing relief when embedded within military initiatives (Hoogvelt, 2007). The murder of 22 UN personnel including its envoy, Sergio Viera de Melo during the suicide bombing of UN Headquarters in Baghdad in 2003 illustrate the risks taken by humanitarian aid workers. Their numbers are rising, and in 2009, 248 relief workers were assaulted or murdered carrying out their work. One of the problems practitioners encounter is not undertaking effective risk assessments regarding their own safety in such situations. Joint endeavours with risk researchers from other disciplines could alleviate such difficulties and make their assessments more reliable.

Disaster responses are complex, controversial and contested. 'Help' may come in the form of internationalising practices that fail to respect local conditions and traditions and become perceived as new forms of colonialism (Mohanty, 2003). For example, the USA has 'tied' aid by requiring that 70 per cent of the sums are spent to employ Americans who will 'help' in disaster relief and purchase goods and services made in America. Roger Riddell (2007) has argued that 'tied' aid restricts opportunities in receiving countries and can distort development opportunities, an issue crucial to long-term reconstruction.

Egalitarian or empowering responses can be problematic to develop, especially if those intervening are not sensitive to local needs, conditions and traditions (Hancock, 1996) or reinforce existing armed conflicts and imperialistic social relations (Hoogvelt, 2007). To complicate matters further, the Geneva Convention does not apply to climate migrants (Sanders, 2009). Responding to these realities requires new protocols (UNDP, 2008). Social workers have an interest in raising these issues through IASSW and IFSW's consultative status in the UN to argue that its mandate: is extended to climate refugees; replaces top-down approaches with collaborative partnerships that engage local residents in solving their own local problems and co-producing new forms of knowledge and approaches to disaster relief; brings together expertise from both physical and social scientists to calculate and assess risk in specific geographical locations and socio-economic, political and cultural contexts and formulate effective, preventative strategies and long-term responses to disaster interventions.

Actions to be undertaken by social work educators, practitioners and policymakers

Social work educators, practitioners and policymakers can participate in disaster intervention debates including climate change and take ethical action by arguing for:

- Human rights and dignity at both individual and group levels.
- Social Justice.
- Interdependence, Reciprocity and Solidarity.
- Peace; and
- Environmental Justice.

Social workers can draw upon ethical principles articulated in the ethics document agreed between IASSW and IFSW in 2004, and in national codes of ethics for practice for working in communities and when collaborating with physical scientists.

Given people's scepticism over the distribution of humanitarian aid (Hennessey, 2009), controversies over its (mis)use (Duffield, 1996, 2007), and governments' failure to reach a binding agreement on climate change in Copenhagen (Booker, 2009, Gray, 2009; Mason, 2009) and Cancun (TWN, 2011), taking action will not be easy. Social workers can become more proactive in policy discussions nationally and internationally by promoting disaster interventions including that of substantially reducing global pollution. Their actions should endorse the following principles:

1 *Interdependence*. We live in an interdependent world where what happens in one location affects people in others and makes us responsible for one another.
2 *Equity*. The inequalities that allow rich or powerful nation-states to dictate terms to low-income ones are unacceptable. Equitable agreements can facilitate transfers of clean technologies to reduce climate change.
3 *Collaboration*. Nation-states collaborating with each other can find solutions to the profound problems raised by disasters including climate change and nuclear meltdown.
4 *Reciprocity*. Reciprocity requires all peoples to contribute to the well-being of others and is essential in recognising interdependence and promoting solidarity.
5 *Solidarity*. Solidarity enables peoples to pool resources, skills and knowledges in addressing problems that humanity faces, regardless of whether an individual personally experiences them.
6 *Sustainability*. Sustainable development on a global scale requires co-ordinated action amongst nation states to share resources and clean

technologies equitably and eliminate inequalities between and within nations, and physical scientists to communicate their knowledge to communities, drawing upon social workers to make these accessible by acting as translators and communications of scientific data.

7 *Holistic approaches*. Each nation-state faces different risks and disasters, e.g., climate change where people contribute differentially to the problem. Current greenhouse-gas emissions will mount as low income countries industrialise to raise people's standards of living. Western lifestyles are physically, spiritually and environmentally unsustainable and cannot be promoted as the way forward for the existing global population, let alone the projected growth beyond 9 billion to be reached by 2030 (UNDP, 2009). Consequently, all nations must work together to find sustainable lifestyles for everyone now living on earth and for the generations to come. This will mean transferring knowledge, skills, and resources more equitably across the world and a holistic approach that encompasses the physical, spiritual and social domains. This requires continued and extensive dialogue between physical and social scientists.

As a profession, social workers can link up with social welfare organisations and other disciplines including engineering, biology, seismology, and geography, to undertake the following actions:

1 *Raise awareness* of the hazards and risks raised by disasters including climate change locally, nationally and globally.
2 *Lobby* the United Nations and national governments for the: affirmation of the above principles; adoption of social policies necessary for their implementation; and commitment to undertaking urgent action to reduce global pollution and over-urbanisation; encourage the equitable sharing of clean technologies and world's resources; and create dignified sustainable lifestyles for all.
3 *Mobilise* people to raise their concerns and promote co-produced solutions to problems at the local, national and international levels.
4 *Research* disasters including climate change and the risks associated with them jointly with physical scientists to: develop new social work perspectives on the issues raised; provide evidence necessary for supporting people in reducing risks before, during and after disasters; and cutting global pollution and its impact on communities locally, nationally and internationally.
5 *Train* people in understanding issues linked to disasters including climate change and ensuring that these are covered in social work curricula. These should include undertaking placements that address disasters and climate change in local communities and joint educational programmes between the physical and social sciences.

Additionally, social workers can strengthen their resilience and contributions to disaster reduction measures by promoting personal, national and international action.

Personal action

Social workers can educate people and raise consciousness about disaster risk reduction and personal resilience strategies. Measures would depend on the nature of anticipated disasters and resources at people's disposal. Existing knowledge in the physical sciences could be instrumental in facilitating this task. For example, those living in flood-plains could 'flood-proof' their homes, e.g., people in Bangladesh have built houses on stilts for the water to flow underneath houses. Individual efforts in climate change are important because domestic emissions comprise 40 per cent of total emissions (Giddens, 2009). Social workers can link up community residents with physical scientists as has occurred in Misa Rumi in Argentina, and the Gilesgate Project initiated by social work academics at Durham University (Dominelli, 2011) to reduce their personal carbon footprint by using less energy, e.g., replacing traditional light-bulbs with energy saving ones, insulating homes, reducing heating by 1°C, limiting reliance on air conditioning, using renewable energy sources like solar panels and heat pumps for heating, switching off electrical gadgets on 'standby' and travelling on public transport. Personal action alone is not enough. Consensual, collective solutions at national and international levels are also required.

National action

National action focuses on empowering local communities in devising solutions to limit damage caused by disasters, using resources wisely and helping those in need. Social workers can mobilise communities to achieve their goals, assist groups to advocate for policies that 'pool risks' and facilitate the co-production of solutions. Local people have good suggestions to make if those making decisions and holding resources would listen to their proposals and include them in their deliberations, as Lane (2011) demonstrated when addressing recurrent flooding problems in North Yorkshire. Social workers can facilitate community involvement in scientific assessments of existing community emergency response plans and involve local people in reformulating these alongside the engineers and emergency planners taking these matters forward.

Transfers of clean energy technologies are part of emission reductions to limit climate change. Funds for successful adaptation and 'green' industrialisation are needed by poor people in low income countries. National governments can facilitate these actions by persuading companies to donate 'green' technologies to low income countries and subsidising them to do so. Governments can fund sustainable 'green' pathways to development that draw upon local strengths and initiatives. Social workers can assist physical scientists to engage with communities in technology transfer initiatives to get ordinary people on board and to reduce waste and avoid duplication. In 2009, the EU estimated that 100 billion euros per year would have to be transferred from rich countries to low-income ones by 2020 for climate change mitigation and adaptation. It suggested that Europe provide $30 billion of this sum yearly; the USA $25 billion; and other industrialised nations, the remainder. These contributions were calculated according to the size of GDP and level of carbon emissions. The EU deemed this affordable because it was about 0.3 per cent of rich countries' annual overall income (*The Week*, 2009, p. 28). Stern (2006) argued addressing climate change concerns now is cheaper than letting them extend into the future. Social workers can use such scientific knowledge to mobilise communities to advocate for carbon reduction initiatives locally, nationally and internationally.

International action

Social work educators and practitioners in IASSW, IFSW and ICSW can develop better collaborative structures to strengthen their involvement in the UN and international agencies to promote actions enhancing people's well-being and their voice at the humanitarian aid table. The framework for international aid in disasters is already in place. IASSW, IFSW and ICSW could embark on more joint activities to advocate for a 'greener' future and advocate for an equitable distribution of resources globally. They can use existing scientific knowledge to advocate for limiting greenhouse emissions to 1,400 billion tonnes between 2000 and 2050 to keep temperature rises below 2°C (Stern, 2006). Such collaboration could also enable physical scientists to become more visionary in transcending the binary paradigm of the 'West as aggressive polluter' and 'Developing world as victim' (Dominelli, 2011) that currently blocks progress on climate change as emerging economies emit significant amounts of greenhouse gases (Dessler and Parsons, 2010). Transcending this reality requires physical scientists to become involved alongside social workers and local residents in co-producing alternative solutions that see the world as one whole and accept interdependency between peoples and countries.

Conclusions

Social workers have important roles to play during times of disaster. They can facilitate working across borders including disciplinary boundaries to engage in mutual knowledge exchanges and transfers with physical scientists including those undertaking risk research; draw upon risk research in the physical sciences to identify disaster-prone areas such as those subject to earthquakes, landslides and flooding and convey such knowledge to local communities; help disaster-proof communities, the physical environment and built infrastructure during the reconstruction and preventative phases of disaster interventions; engage multiple stakeholders including policymakers in co-producing knowledge and solutions to advance and innovate beyond current possibilities; involve physical scientists in community initiatives including securing their commitment to working within a human-rights, social justice framework which will bring the politics of risk research into the centre of scientific endeavours rather than leaving these at the margins. Through such collaborative ventures, social workers can become more scientific and credible in the roles they adopt when:

- Delivering humanitarian aid.
- Long-term reconstruction.
- Developing risk reduction strategies and action plans.
- Advocating locally, nationally and internationally for the equitable distribution of resources and 'green' technologies.
- Educating and raising consciousness of preventative measures in disaster-prone areas;.
- Strengthening community groups' roles in working with physical scientists to co-produce solutions; and
- Changing social work curricula to train practitioners in working across the physical and social science divide.

Throughout these activities, social workers will also have to act ethically and will be held accountable for their behaviour during disaster responses. The same will hold for physical scientists and risk researchers engaging with those living in disaster-prone communities.

Additionally, social workers and physical scientists can work together to:

- Create a new agenda for risk research.
- Develop new risk assessment tools for practitioners.
- Integrate complexity and human agency in disaster intervention models.
- Develop ethical guidelines to be followed during all phases of disaster responses by all those involved in supporting vulnerable communities.

References

Askeland, G. (2007) Globalisation and a Flood of Travellers: Flooded Travellers and Social Justice in *Revitalising Communities in a Globalising World*, (ed L. Dominelli), Ashgate, Aldershot.

Austin, L. (1992) *Responding to Disasters: A Guideline for Mental Health Professionals*, American Psychiatric Publishing Inc, Arlington, VA.

Bragin, M. (2008) *No Harm UN International Guidelines on Mental Health and Psycho-Social Support in Emergency Settings.* Paper given at the IASSW-KKI Disaster Reduction Conference in Durban, South Africa, 25 July.

Bywaters, P. (2008) Learning from Experience: Developing a Research Strategy for Social Work in the UK. *British Journal of Social Work*, **38**(5), 936–952.

Correll, D. (2008) The Politics of Poverty and Social Development. *International Social Work*, **51**(4), 453–466.

Desai, A. (2007) Disaster and Social Work Responses, in *Revitalising Communities in a Globalising World* (ed L. Dominelli), Ashgate, Aldershot.

Dessler, A. and Parsons, E. (2010) *The Science and Politics of Global Climate Change: A Guide to the Debate*, Cambridge University Press, Cambridge.

Dominelli, L. (1998) in Adams, R., Dominelli, L. and Payne, M. (eds) Anti-oppressive practice in context. In Social Work: Themes, Issues and Critical Debates. Houndmills: Palgrave Macmillan, 416pp.

Dominelli, L. (2004) *Social Work: Theory and Practice in a Changing Profession*, Polity Press, Cambridge.

Dominelli, L. (ed) (2007) *Revitalising Communities in a Globalising World*, Ashgate, Aldershot.

Dominelli, L. (2009) Social Work Research: Contested Knowledge for Practice, in *Practising Social Work in a Complex World* (eds R. Adams, L. Dominelli and M. Payne), Palgrave Macmillan. Second edition. First published in 2005, London.

Dominelli, L. (2009) *Introducing Social Work*, Polity Press, Cambridge.

Dominelli, L. (2010) *Social Work in a Globalising World*, Polity Press, Cambridge.

Dominelli, L. (2011) Climate Change: A Social Work Perspective. *International Journal of Social Welfare*, forthcoming.

Duffield, M. (1996) The Symphony of the Damned: Racial Discourse, Complex Political Emergencies and Humanitarian Aid. *Disasters*, **20**(3), 173–193.

Duffield, M. (2007) *Development, Security and Unending War*, Polity Press, Cambridge.

Giddens, A. (2009) *The Politics of Climate Change*, Polity Press, Cambridge.

Gunn, S.-W.-A. (1990) *Multilingual Dictionary of Disaster Medicine and International Relief*, Kluwer Academic Publishers, Boston.

Hancock, J. (1996) *The Lords of Poverty: The Power, Prestige, and Corruption of the Interntional Aid Business*, Atlanta Monthly Press (First ed in 1989), New York.

Hoogvelt, A. (2007) Globalisation and Imperialism: Wars and Humanitarian Intervention in *Revitalising Communities in a Globalising World* (ed L. Dominelli), Ashgate, Aldershot.

IFRC (2009) *World Disasters Report, 2009*, IFRC, Geneva.

IPCC (Intergovernmental Panel on Climate Change) (2007) *The Fourth Assessment on Climate Change*, IPCC, New York.

Javadian, R. (2007) Social Work Responses to Earthquake Disasters: A Social Work Intervention in Bam, Iran', *International Social Work*, **50**(3), 334–346.

Kassindja, F. and Miller-Basher, L. (1998) *Do They Hear You When You Cry?*, Delacourt Press, New York.

Laming, H. (2009) *The Protection of Children in England: A Progress Report*, DCSF, London.

Lane, S. (2011) Doing Flood Risks Differently: An Experiment in Radical Scientific Method. *Transactions of the Institute of British Geographers*, **36**(1), 15–36.

Li, Y., Huang, R., Yau, L., Densmore, A. and Zhou, F. (2010) Surface rupture and Hazard of Wenchauan Earthquake. *International Journal of Geosciences*, **1**(1), 21–31.

Loescher, G., Betts, A. and Milner, J. (2008) *UNHCR: The Politics and Practice of Refugee Protection into the Twenty-First Century*, Routledge, London.

Manyena, S. B. (2006) The Concept of Resilience Revisited. *Disasters*, **30**(4), 434–450.

Mohanty, C. T. (2003) *Feminism Without Borders: Decolonizing Theory, Practicing Solidarity*, Duke University Press, London.

Perez, E. and Thompson, P. (1994) Natural Hazards: Causes and Effects. *Pre-hospital Disaster Medicine*, **9**(1), 80–88.

Pittaway, E., Bartolomei, L. and Rees, S. (2007) Gendered Dimensions of the 2004 Tsunami and a Potential Social Work Response in Post-Disaster Situations. *International Social Work*, **50**(3), 295–306.

Parton, N. (2004) From Maria Colwell to Victoria Climbié. *Child Abuse Review*, **13**(2), 80–94.

Pattison, G. (2007) Community Participation: A Critical Appraisal of the Role of 'Community' in Urban Policy, in *Revitalising Communities in a Globalising World* (ed L. Dominelli), Ashgate, Aldershot.

Pearlstein, D. and Posner, M. (2009) Brief in Support of S A Hamadan v D H Rusmfeld. At http://www.humanrightsfirst.org/us_law/PDF/hamdan_v_Rumsfeld_Brief.pdf accessed 29 Dec 09.

Popple, K. (2007) Community Development Strategies in the UK, in *Revitalising Communities in a Globalising World* (ed L. Dominelli), Ashgate, Aldershot.

Pyles, L. (2007) Community Organising for Post-Disaster Development: Locating Social Work. *International Social Work*, **50**(3), 321–333.

Quinsey, V. L. (1995) Predicting Sexual Offences: Assessing Dangerousness, in *Violence by Sexual Offenders: Batterers and Child Abusers* (ed J. C. Campbell), Sage, London.

Ramon, S. (2008) (ed) *Social Work in the Context of Political Conflict*, Venture Press, Birmingham.

Riddell, R. (2007) *Does Aid Really Work?* Oxford University Press, Oxford.

Sanders, E. (2009) Climate Change Creates Refugees, in *The Vancouver Sun*, 26 October, p. B5.

Shiva, V. (2003) Food Rights, Free Trade and Fascism' in *Globalizing Rights* (ed M. J. Gibney), Oxford University, Oxford.

Steiner, N. (2003) *Problems of Protection*, Routledge, London.

Stern, N. (2006) *Stern Review of the Economics of Climate Change*. Cambridge University Press, Cambridge.

Swift, K. and Callahan, M. (2010) *At Risk: Social Justice in Child Welfare and Other Human Service*, Toronto University Press, Toronto.

Terry, F. (2002) *Condemned to Repeat? The Paradox of Humanitarian Action*, Cornell University Press, Ithaca, NY.

Tramonte, M. (2001) *Risk Prevention for All Children and Adults: Lessons Learned from Disaster Interventions*, NCMPV, Newton.

TWN (Third World Network) (2010) Cancun Briefing Papers, 29 November – 10 December on www.twnside.org accessed 29 Dec 2010.

UNDP (2007) *Fighting Climate Change: Human Solidarity in a Divided World*, Palgrave/Macmillan, London.

UNDP (2008) *Climate Change: Scaling Up to Meet the Challenge*, UNDP, New York.

UNDP (2009) *The Human Development Report*, UNDP, New York.

UNFCCC (2008) *Disaster Risk Reduction Strategy and Risk Management Practices: Critical Elements for Adjusting to Climate Change*.
Yule, W. (2008) IASC Guidelines, Generally Welcome, But..., *Intervention*, **6**(3), 248–251.

Useful websites

www.un.org
www.undp.org
www.unhrc.org
www.unhcr.org

CHAPTER 11

Conclusion: Reflections on 'Critical' Risk Research

Stuart N. Lane[1], *Francisco R. Klauser*[2] *& Matthew B. Kearnes*[3]

[1]Institut de Géographie, Faculté des Geosciences et de l'Environnement, Université de Lausanne, Lausanne, Switzerland

[2]Institut de Géographie, Faculté de Lettres et Sciences Humaines, Université de Neuchâtel, Neuchâtel, Switzerland

[3]School of History and Philosophy, University of New South Wales, Australia

The chapters collected in this volume display an extraordinary range of conceptual reflection, across a range of spatially and temporally distributed sites of empirical investigation. Given this heterogeneity what is perhaps most remarkable is the way in which a cluster of themes around the notion of criticality – and indeed need for critical interventions in the practice of risk research, risk assessment and post-disaster relief efforts – emerge across these chapters. At issue here is a shared concern for what might be thought of as *situatedness* of critical risk research; with the social, economic and political contexts that shape the conduct of approaches to risk, the continuing and persistent influence of particular academic disciplines at the expense of others, and the provisionality of expert knowledge claims (Fischer, 1999; Hajer, 1995; Jasanoff, 1986). In addition, the chapters that comprise this volume highlight the ethical and normative implications of the situatedness of contemporary risk research, arguing that an important first step in generating self-reflexive approaches requires subjecting both the epistemological constitution and practical conduct of risk research to critical scrutiny.

In introducing this volume, we identified three core concerns – conceptualisation, disciplinarity and interdisciplinarity, and institutionalisation – as critical for contemporary risk research. In many ways these three themes have, in recent years, become the preeminent conceptual and practical challenges for risk researchers; challenges that are revealed to be both pressing and salient in the aftermath of events such as the 2004 Indian Ocean earthquake and tsunami, as well as the 2011 Tōhoku earthquake in Japan. In response to these challenges across the chapters that of this

Critical Risk Research: Practices, Politics and Ethics, First Edition.
Edited by Matthew Kearnes, Francisco Klauser and Stuart Lane.

volume we identify a range of cross-cutting and thematic concerns and observations that speak to the often ambivalent relationships researchers maintain with the social and political factors that form the context of contemporary risk research. Across each of the chapters concern centres on how risk as a concept is put to work; the importance of reflecting upon how risk is made calculable; and, crucially, the need to rethink notions of expertise and authority, which are typically implicit to the conduct of risk research and officially sanctioned forms of risk assessment and management. In addition to these issues a range of additional, more synthetic, themes are evident across the chapters. These include a concern for the increasing institutionalisation of risk research, particularly as a technology of state-power (Amoore and de Goede, 2008; Collier, 2008; Ericson, 2005; O'Malley, 2004), the challenge posed by interdisciplinary ways of working (Bracken and Oughton, 2009) and the development of new forms of public participation and deliberation (Chilvers, 2008, 2010; Lane *et al.*, 2011). In summarising and reflecting on these themes we conclude by suggesting that if risk researchers are to respond to the social and political contexts that shape research in the field this will, above all, require nurturing a set of new ethical and normative capacities – a capacity for reflexive and critical research practice that enables the field to become more porous, open to a range of non-traditional and lay knowledges and alternative forms of expertise grounded in the everyday perceptions and experiences of risk.

Putting risk to work

The first underlying theme, evident across the chapters of this volume, is that risk is a concept that is 'put to work', one that serves to maintain existing social and political orders and, in so doing, takes on an existential logic. Risks are created and constructed from the panoply of possible threats affecting a population, society or social group (Adam *et al.*, 2000). By translating both known and anticipated threats into risks, and by enabling the development of new institutional and governmental forms, risk research and risk management become a technology of contemporary political power. For example, in their study of the development and practices of surveillance at Geneva International Airport, Klauser and Ruegg (this volume) demonstrate how these security and policing systems were established not simply as a necessary response to security threats. Rather, the development of these systems was also informed by the commercial appeal of the airport. The case exemplifies that critical risk research and management must always be positioned within a complex field of both often conflictual but mutually reinforcing public and private imperatives, driving forces and motivations. The goals and strategies of risk management are constantly renegotiated and reoriented within this complex network of expertise and authority.

Often with little oversight or official sanction, the techniques and tools of risk management and securitisation are utilised in ways well beyond their initial function. The terminology of risk, threat and vulnerability becomes a technically inscribed vocabulary for a whole host of contemporary problems and problematics.[1] As the chapters of this volume suggest, risk is put to use in at least two principle ways – by conferring a form of technical legitimacy on a range of social and political discourses and by constituting an underlying technology of contemporary government. It is this distinction, between what Rose and Miller (1992) term 'political rationalities' and the 'technologies of government' that defines the intersections between risk and contemporary political power. Rose and Miller outline this distinction in thc following way, suggesting that:

> It is through technologies that political rationalities and the programmes of government that articulate them become capable of deployment. But this is not a matter of the 'Implementation' of ideal schemes in the real, nor of the extension of control from the seat of power into the minutiae of existence. Rather, it is a question of the complex assemblage of diverse forces – legal, architectural, professional, administrative, financial, judgmental – such that aspects of the decisions and actions of individuals, groups, organisations and populations come to be understood and regulated in relation to authoritative criteria. (p. 183)

In this volume we have seen that the concept of risk operates on both sides of this distinction. On the one hand, risk embodies a certain kind of political rationality or logic, where the regulation and management of risky things and risky bodies becomes central to the imagination of contemporary government (O'Malley, 2004). On the other hand, institutionally-sanctioned forms of risk management represent one of 'the humble and mundane mechanisms by which authorities seek to instantiate government' (Rose and Miller, 1992, 183).

Taken as a whole, the question that is posed by this volume concerns the implications of this co-production of risk and contemporary political

---

[1] This comment strongly echoes Winner's conception of the 'function creep' of technology (Winner, 1977) See also Lyon (2007) and (Haggerty and Ericson, 2000).

order for critical risk research and risk researchers. A number of chapters in this volume have pointed to the ethics of risk research, demonstrating that risk problems are bound, both directly and indirectly, to a range of ethical and normative concerns (Lane, this volume; Macnaghten and Chilvers, this volume; Rigg *et al.* this volume). However, as demonstrated by these interventions, the relationship between ethics and risk is not straightforward. Through risk research clearly has an 'ethical dimension', worthy of dedicated and thoughtful reflection the political utility of the concept of risk is also sustained by a 'risk industry', which provides a context of contemporary risk research (see Klauser and Ruegg, this volume; Macnaghten and Chilvers, this volume). As Kearnes (this volume) suggests, notions of virtue, responsibility and ethics are increasingly invoked in the constitution of contemporary risk analysis. Public displays of corporate responsibility, political transparency and accountability have become the stuff of institutionalised forms of risk management, often called or recalled so as to resuscitate the social and political legitimacy of governmental interventions (Brown and Michael, 2002; Shamir, 2010; Strathern, 2005). In this sense, critical risk research is ethically significant in at least two ways. Whilst risk management and security interventions evolve in ways that have important (ethically relevant) power implications (Klauser and Ruegg, this volume), these interventions, including persistent calls for greater and more comprehensive forms of surveillance and risk management, become justified as moral imperatives in themselves. By designating threats as risks on ethical grounds it becomes possible to justify interventions whilst simultaneously overlooking the cultural and political forces that motivate, and which rely upon, this designation. Risk is put to work through its construction as an 'ethical' practice, notwithstanding the much less ethical behaviours that are then sustained by this construction.

Second, when risk is manifest as a (catastrophic) event, that event may itself, in turn, become an opportunity for critical risk research. Davies *et al.* (this volume) opened their innovative chapter (a narrative between three academics) by noting just how much we rely upon the definition of a problem as 'risky' so as to justify the research related to that problem. This is almost 'risk' without limits: if we define risk using one of the classical formulae we noted in the Introduction, where risk is simply calculated as a product of exposure and probability, all things have the potential to be mobilised as risk. What Davies *et al.*'s chapter points to is the malleability of risk, which is made all the more effective as a foundation for academic enquiry because of its moral imperative. Ethically motivated risk analysis and management may share strong parallels with ethically motivated risk research.

Making risk calculable

A related theme, across a number of chapters in this volume, concerns issues of calculation and quantification. Much like the man with the hammer who sees every problem in the form of a nail, standardised mechanisms for the calculation and quantification of threats represent the managerial and techno-scientific *zeitgeist* of current approaches to dealing with risks. In this sense, perhaps the primary way in which risk is put to use in sustaining the contours of contemporary political power is rendering the uncertain and threatening calculable. For example Dean (1999) suggests that 'risk is a way – or rather, a set of different ways – of ordering reality, of rendering it into a calculable form. It is a way of representing events so they might be made governable in particular ways, with particular techniques, and for particular goals' (p. 131). Notions of calculation and calculability therefore define the epistemological limits of the risk, such that 'it is thus not possible to speak of incalculable risks, or of risks that escape our modes of calculation, and even less to speak of a social order in which risk is largely calculable and contrast it with one in which risk has become largely incalculable' (*ibid*, p. 131). The primary utility of the concept of risk – and the way that it is invested with a range of contemporary anxieties about our relationship with the natural world, with other people and populations and with the fruits of our technological accomplishments – is to render dimly conceived threats and hybrid materialities knowable and so to justify the investment of resources in subsequent (risk) management. Risk is a culturally functional term that serves to make the anticipated, the possible and the predicted socially tractable. It is the vocabulary through which particular threats can be made 'public problems' (Borraz, 2007; Gusfield, 1981).

In practice, however, the chapters of this volume demonstrate a range of ambivalences in this intersection between the concept of risk and notions of calculation and quantification. Principle here is a concern for the ways in which standardised mechanisms for the measurement and quantification of risk are underpinned by and reinforce a limiting set of framings. These include: the expression of risky activities in terms of costs and benefits; the quantification of the probabilities of loss whether of life, money or other 'resources'; the use of indices of health and well-being as a metric of both impact and performance. As we noted in the Introduction such framings themselves are in need of critical analysis. Rigg *et al.* (this volume) for instance challenges the ways in which the calculation and assessment of risks becomes spatially and temporally delimited, typically centring on geographically, and historically, defined communities and where elucidating impacts of events becomes possible. Rather, and in line with our comments in the Introduction regarding the issues of scale as revealed by events such as the Fukushima catastrophe, Rigg *et al.* demonstrate how the implications of such events need to be viewed in terms of a much wider network, where the impacts are recognised as having effects in a more geographically distributed and variegated pattern, expending well beyond the location of the physical 'event'.[2] The issue here concerns the geographical and historical fixity of contemporary risk research, and the ways in which current approaches, by relying in notions of the physical *calculability* and quantification of risk, are implicitly bound in both time and space. As Rigg *et al.* demonstrate, the danger here is that these approaches produce partial accounts of risks and vulnerabilities that effectively deny the felt experience of affected populations. The challenge for critical risk research is to avoid framings that are explicitly bounded in space and time even if this implies a tension with the need for critical risk research that is attentive to the geographical and historical specificity of contemporary risks.

Making risk calculable also involves a second tension. Risk is often seen as the 'industry of uncertainty'. Take, for example, the reinsurance industry: it is built around the implicit assumption that whilst risk is ever present, events are geographically and historically random. For most of the time, most communities and individuals are free from such events, but may be cognisant of the presence of risk. This cognisance may be expressed financially through the payment of insurance premiums, by the many, and capital accumulates in the risk industry as long as risk does not translate into events. When it does, capital is released to those geographically less-extensive places, those deemed to have been effected by events. The fact that risk is uncertain is what allows this model to work: present and with a randomness that is quantifiable; aleatory. However, not all uncertainty is aleatory and making risk calculable is, therefore, about reducing that which is uncertain to that which is aleatory (Stirling, 2009). To be made calculable, 'risky things' need to be reduced to their physical essence, described by a finite set of possible rules, whether those are deterministic or based upon the stochastic characteristics of past events. Indeed, as developed by Kearnes (this volume), risk management has tended to see itself as the arbiter of the *physical* properties of risky things, an arbitration that becomes sustained by keeping non-physical elements out of the analysis. Risk becomes governable because in this stripped down state, priorities can be set on the grounds of apparently objective criteria, and places can be compared in terms of their 'riskiness'. The lifeworld of the

---

[2] See Anderson (2010) for a discussion on an expanded definition of 'the event' of revelance to an extended notion of risk, impact and harm.

risk analyst is comparative, one in which history and geography become atomised into comparable return periods. Not only, as Lane (this volume) argues, are such 'atoms' far removed from how risk is experienced – in his case by flood victims – but the atomisation is equally founded upon a series of assumptions and simplifications that mean that the atoms can only continue to exist if the world that has shaped their constitution is, in turn, made to look like them. Risk calculation does not tell us about risk in the world, as we know it. Rather, risk calculation tells us what the world has to be made to look like, if those calculations are indeed to become correct (Lane *et al.*, 2011). This is an example of the co-production of risk analysis by risk management and of risk management by risk analysis.

Calculating risk, in this way, leads to an important consequence. Not only will events with low return periods still occur (because they have a longer return period than the standard of protection that we have chosen) but they are likely to happen much more frequently than we might imagine because we cannot make the world look like our calculations of it. Events, in a risky world, should expected to be more frequent and potentially more catastrophic than our calculations suggest. The Fukushima event discussed in the Introduction provides a terrible example thereof. Such events force us to interrogate those calculations, and the institutions behind them (Macnaghten and Chilvers, this volume), not only to reveal their bias, but also to open up the possibility for that which has not, or cannot, be quantified to be brought back into the frame.

A third tension then arises. We tend to see risk research as the analytically determinate and independent arbiter of risk management priorities. In a classically realist frame the purpose of critical risk research is to uncover the true nature of possible and anticipated threats, and to provide a knowledge base for possible remedial actions. However, it is also clear that in the act of managing risk, risk itself becomes redefined, and risk research is actively involved in this redefinition. For instance, Klauser and Ruegg (this volume) describe how it is technological innovation in security companies that influences how risk is revealed and which, in turn, causes risk to be constructed in new ways. As with other areas of both the natural and social sciences, the act of researching risk is not simply neutral, but contributes directly and indirectly to the reformulation of its objects of enquiry. Risk research is actively involved in the co-production of risk analysis, risk management and of itself.

The chapters of this book suggest at least three possible strategies, or areas for further reflection, to respond to these tensions. First, there is a need to think through who and what makes risk calculable and, in particular, to consider the range of social, political and economic interests at play in rendering some threats as quantifiable risks. Such an analysis needs, equally, to be dynamic by recognising that the designation and conception of particular risks is one that may equally be driven by experience of the act of calculation as much as perceived threats themselves. Secondly, while non-physical elements have been implicated in our understanding of risk for some time (see for example, Douglas and Wildavsky, 1982) the chapters of this book demonstrate that what is much less well thought through is precisely how it is that critical risk research has grasped the challenge of understanding risk as a hybrid of nature and culture. Most pointedly the chapters demonstrate how the academy has worked with and against this hybridity (Beck, 1998). For example, both Bracken (this volume) and Rigg *et al.* (this volume) reflect upon the consequences of disciplinary lock-in for situations that are clearly hybrid, and where interdisciplinary modes of analysis and engagement are required. Both highlight the subtle ways in which particular disciplinary traditions come to the fore in the definitions of particular risks and hazards, defining research problems in specific ways. Third, the recognition of risk as hybrid is potentially valuable. It points in a normative sense to what risk analysis and management ought to involve. But, it also indicates how critical risk research needs to respond by experimenting with new ways in which risk analysis and management might work with this hybridity rather than advocating its reduction, atomistically, to non-physical things. It is perhaps interesting that many of the chapters in this volume (Bracken, Davies *et al.*, Dominelli, Kearnes, Lane, Macnaghten and Chivers, Rigg *et al.*) have considered the need for new forms of research into democratic engagement in the analysis and management of risk, ones which can unsettle the dominant framings that are brought to bear on risk questions, such as the emphasis on comparative prioritisation of risky locations for subsequent intervention.

The state, institutions and governance

An additional area of concern across a number of the chapters here is the need to critically examine the role of institutions, both public and private, in making risks calculable (e.g. Kearnes; Klauser and Ruegg; Lane; Macnaghten and Chilvers, this volume). Nowhere is the interpenetration between risk and social, political and economic order more palpable than in the institutionally sanctioned ways in which some threats are designated as risks. For example, Lane (this volume) argues that the history of risk management might be seen as a progressive change in scale of responsibility for managing risk, from smaller scales to larger scales, from individuals and communities, to institutions and states. In addition he also argues that risk modernisation has been as much about a changing scale and increasing appropriation of institutional power as it has been the progress of technological solutions. The state starts with some form of moral or ethical imperative, perhaps linked to its democratic foundations. Threats become an opportunity to express this imperative, especially if they are manifest as events, and this expression takes the form of interventions and the institutions needed to sustain them. Such institutions usurp authority, drawing further upon moral and ethical imperatives, and so are able to sustain their own logic, their own existence. Events are particularly interesting in this context because they are at once a problem (the possibility that risk as manifest is taken as a sign of institutional failing) and a benefit (an opportunity to further assimilate power into an institution), provided, critically, the event can be recast as a problem that is beyond an institution's control rather than being a sign of institutional failure. As Lane (this volume) notes, research has shown that institutions that are tasked with responsibilities for risk management do not limit their interventions solely to addressing the possibilities of physical harm. Rather such bodies typically engage in a range of 'public relations' exercises that are as much about managing institutional reputations and political positionalities as they are about attenuating the effects of hazardous events. Seen in this light, such events only continue to have their logic in so far as they can be put to work (Collier, 2008; Cooper, 2010).

As part of this transition, Kearnes (this volume) describes an 'anticipatory turn' in contemporary risk analysis and management, which is linked to the development of an institutional capacity to attenuate the latent potentiality of future threats. Traditionally, the governance of risk has been concerned with precaution, with recognising that where the risk is thought to be high, moving forward should be reflective, evaluative and adaptive. Requiring a kind of perpetual recalculation this form of risk governance has led to a strong and mutually reinforcing relationship between what Kearnes called the 'technical logics of prevention and prediction' and the 'moral virtues of precaution and care'. Precaution has a moral imperative that binds it to recalculation. Kearnes argues that risk analysis and management has moved on from this model, in an environment where moving forward meant being prepared to be wrong, to an environment where being wrong is not possible. This may well be sustained by the progressive accumulation of responsibility for risk analysis and management in the state and its institutions (Lane, this volume) and at least where democracy permits it, perceived institutional failures can be addressed through traditional political means.

Extending this theme, Macnaghten and Chilvers' (this volume) analysis of the range of public engagement initiatives, recently commissioned by the UK government, reveals that public mistrust in state institutions has become a constitutional condition for contemporary risk management (see also Hagendijk and Irwin, 2006; Wynne, 2006). Here, the need for recalculation is not bypassed, but shifted; away from calculating the risk towards calculating the conditions that might lead to the risk being manifest. Such calculations have, in turn, invoked new kinds of anticipatory technologies and forms of automated surveillance and management (Amoore and de Goede, 2005; Graham and Wood, 2003; O'Malley, 2010). Similarly, Klauser and Ruegg (this volume) also describe how the ways in which anticipatory risk management strategies – engaged managing the possibilities of security breaches at airports – is embodied in particular kinds of material objects (e.g. closed-circuit television systems). They point to a circular logic whereby these interventions create new experiences of what constitutes risk (e.g. the homeless) and hence what needs to be anticipated. The result is a culture of dependency on a surveillance system that is an evolving hybrid of technology, expertise and experience. The ways in which surveillance systems and risk management processes are augmented by technical devices further suggests their purpose is not only simply the reduction and attenuation of risks but is also part of a social and political assemblage that works to enforce a set of hidden norms about what constitutes acceptable behaviour. The risk begets anticipation and anticipation in turn begets the risk. Surveillance is one of those activities that have become increasingly endemic in the risk industry and it is clear that critical risk research is only just starting to think through what it means for the dynamic of risk and the ethical concerns that are bound to it.

However, whilst calculation, or recalculation may still be endemic to the anticipatory turn, both Kearnes (this volume) and Macnaghten and Chivers (this volume) argue that it also provides an opportunity for rethinking the nature of risk analysis and management. It suggests perhaps a deeper unease with precaution and its underlying faith in the ability to make risk calculable. Kearnes (this volume) argues that it opens up a much richer form of risk analysis and management, one that effectively moves 'upstream', away from *post hoc* in situations where things are discovered to have gone wrong (or risky things are innocent until proven guilty) towards *ante hoc* analyses in the form of a pre-emptive logic (or risky things are seen as forever potentially guilty). In relation to new technologies, in their respective chapters both Kearnes (this volume) and Macnaghten and Chilvers (this volume) argue that such *ante hoc* analyses recognise explicitly that some things (e.g. 'futures') are by definition not determinate. In turn, the upstream move opens up new opportunities, both for a broader evaluation of those involved in producing novel risks (e.g. the financial motives of biotechnology companies) but also for new ways of defining and manipulating threats as risks in ways that sustain and enable the range of actors and institutions bound to them.

This pre-emptive logic is also interesting because of the contradiction that it represents for institutions. On the one hand, pre-emptive risk management is typically presented as more efficient and cost-effective than clearing up afterwards, especially clearing up after perceived institutional failures. For example, areas of technological innovation as diverse as nanotechnology, geoengineering and neuro-chemistry are presented as an opportunity to avoid the mistakes of the past – to circumvent the often reluctant admissions of institutional failures in the face of public concern and political controversy that characterise debates about civil nuclear power and genetically modified food – by engaging in early forms of public participation and by designing anticipatory and adaptive forms of governance and regulation (Barben *et al.*, 2008; Royal Commission on Environmental Pollution, 2008). However, the contradiction is that effective pre-emption rarely becomes appreciated as such, without the aid of a host of devices for simulating the effects of non-intervention. This takes us back to the problematic nature of events, of how risky landscapes have become sanitised of risk by pre-emption and what this means for how we come to live with risk. Concurrently, Kearnes (this volume) shows how the engagement of anticipatory forms of risk management, particularly by state institutions, creates a tension between the capability of the state to know and the capability of the state to act. Ultimately, this leads to an aggrandisement of power where the state claims a monopoly in regulatory expertise in order to pre-empt potential future threats. On the basis of this monopoly a range of additional measures are often engaged to translate anticipatory risk assessment into broadly defined norms of personal responsibility.[3] In this context experiential know-how and lay knowledge is often replaced by de-contextualised notions of expertise invoked in anticipated particular problems and the definition of acceptable modes of responsible behaviour. The state and institutions anticipate, people are expected to act and the dependency on systems of governance, noted above, is reinforced. In turn this rests upon not just the moral authority of the state to engender response but also faith in those instruments of state, such as institutions, engaged in the anticipation of threats and the deeper set of relationships between potential victims, institutions and the state, as they play out in the world around them.

Interdisciplinarity

Whilst chapters in this book raise some crucial new questions about how risk research can illuminate the changing nature of risk analysis and

---

[3] This translation of anticipatory risk management into notions of personalised responsibility and prudence is particularly evident in cases such as the publication of online flood maps (Landström *et al.*, 2011), norms of parental responsibility invoked around childhood obesity (Colls and Evans, 2008) and notions of bodily comportment invoked in strategies designed to address recent outbreaks of influenza (Diprose *et al.*, 2008; Lakoff and Collier, 2008).

management, they equally raise interesting questions about the very nature of critical risk research itself. Davies *et al.* (this volume) draws attention to the power that risk has as a means of legitimating more than just state action, but research itself. If the changing nature of state involvement in questions regarding risk merits consideration as a political economy, so the relationship between disciplines, disciplinary claims regarding critical risk research and the nature of risk also requires interrogation. Both Bracken (this volume) and Rigg *et al.* (this volume) consider how particular disciplinary traditions structure the conduct of risk research, framing the range of knowledge and expertise deemed relevant in addressing particular problems – a parallel between how institutions frame risk and how disciplines frame risk research, in turn shaping how risk becomes understood. It is perhaps interesting that across the chapters of this volume at least three disciplines are positioned as central to the conduct of critical risk research: Anthropology (Merli); Geography (Rigg *et al.*); and Social Work (Dominelli *et al.*). However, each of these chapters reveals that in practice these disciplinary traditions function to render particular explanations of risk dominant and that specific kinds of interventions, both before, during and after events, are legitimated over others. There is a clear need for both comparative analysis of those involved in risk related research projects (Bracken, this volume) but also more specific self-interrogation of the practices bound to risk research as it unfolds (Rigg *et al.*, this volume). Bracken's chapter points to a set of factors that come together to make the practice of interdisciplinary risk research difficult. However Bracken also notes the apparent moral imperative in contemporary risk research, to make a positive contribution to the world around them. Such imperatives appear to be legitimating devices in relation to both the decision to research risk and to translate threats into risks. Perhaps reflecting the complex multitude of ways in which risk may be defined, and the consequences of this definition for the kinds of risk management that result, interdisciplinarity and the normalisation of interdisciplinarity as a practice are invoked as necessary to making risk research 'better' (Rigg *et al.*, this volume). In theory, interdisciplinarity can unsettle the framings brought by particular disciplines. However, both Bracken (this volume) and Rigg *et al.* (this volume) demonstrate that the ideal of interdisciplinary critical risk research partly founders on the power of disciplines, on their political economy.

A number of chapters in this volume point to ways in which these framings can become unsettled in a way that results in openings for the inclusion of alternative forms of disciplinary explanations. In the same way that events can unsettle institutions, so they may also unsettle the nature of risk research forcing interdisciplinarity. But as Rigg *et al.* (this volume) reflected, whilst this may allow the development of new understandings of particular events, how far this understanding can travel back into the academic world can become strongly conditioned by disciplinary forces. If events are not enough, risk researchers need to develop other ways of letting the essence of risk 'speak back' (Bracken *et al.*, this volume), involving both animate and inanimate objects (Kearnes, this volume).

People and participation

The final theme common across the chapters of this volume is a response to this problem – that recognising the 'moral imperative' of critical risk research this requires new forms of participatory risk research, principally by engaging with those who live with, and are affected by, contemporary risks (see Dominelli, this volume, Rigg *et al.*, this volume). Lane (this volume) reflects upon how, in his research, working with those who live with floods forced him to dissociate with his normal networks of academic enquiry and to associate with those who had been excluded from the practice of generating knowledge. He found strong parallels between the knowledge gained through the extensive and comparative practices of the certified expert and the local and place-bound expertise of the disaster victim; but also a strong difference, one echoed strongly by Merli (this volume). The knowledge of the (certified) expert is knowledge commonly gained through a professional position, which may in turn invoke a personal attachment to that problem; but the non-certified expert gains knowledge through lived experience, where the nature of that experience and the motivations it creates are fundamentally different. There are very few (certified) experts in relation to disasters if knowledge is defined as 'lived experience'. For example, Merli argues that, if risk research started with definitions of risk, threat and harm as experienced by non-certified experts this would have the effect of fundamentally changing the ways in which these risks are constructed and understood by regulatory agencies and institutions.

The call for more engaging forms of public participation is then a means of unsettling dominant framings in risk analysis and management. For instance, institutions commonly bemoan the differential between public risk perceptions and the 'reality' of risks and the consequent failure of populations to act rationally when facing risks or to head official warnings and advice. Wynne (1996, 2001) characterises such a view – that fickle and irrational public perceptions need to be addressed simply through clearer and more comprehensive risk communication – as the 'deficit model' of the public understanding of science. He demonstrates that across a series of recent controversies a guiding assumption on the part of regulators has tended to be that public concerns are the result of collective irrationality, caused by a deficit in public knowledge of the scientific 'reality' of risks. Notions of public irrationality, and that public concerns are often inflamed by popular media reporting, have become a kind of folk-story deployed in explaining the emergence and persistence of public concerns in the face of regulatory assurances.

However, Wynne (2006) goes on to demonstrate that, in many cases, programmes that have aimed to improve public understanding of risk issues and general scientific literacy have often had the inverse effect, working to intensify underlying concerns and anxieties. The chapters in this volume corroborate this finding, suggesting that the often incompatible relationship between public risk perceptions and institutionally-authorised expert judgement is not simply based on a knowledge differential or that researchers should work to overcome 'local' traditions and belief systems in the conduct of critical risk research and communication initiatives (Merli *et al.*, this volume; Rigg *et al.*, this volume). Rather this incompatibility is produced by an ontological distinction (Verran, 2001) that conditions how different social groups respond to and manage risks and catastrophic events. In addition, recent psychological research has established for some time the complex playing out of risk and reward, and their subsequent path dependence (see Lane, this volume; Rigg *et al.*, this volume). Rather than events being mapped as the random manifestation of a statistically stationary magnitude-frequency relationship, events have to be seen as emergent as they are created by (and also transform) the complex assemblage of which they are a part. Risk then, is produced through the way it is lived, and risk itself is a lived experience. What is at stake in contemporary critical risk research is what Mol (2002) describes as 'ontological politics'. She suggests that 'ontology is not given in the order of things . . . instead, ontologies are brought into being, sustained, or allowed to wither away in common, day-to-day, socio-material practices' (p. 6). In a similar fashion, Dominelli, Merli and Rigg *et al.* (this volume), draw out the ontological diversity of risk as it is both experienced and practiced. If this is how risk is experienced, critical risk research needs to be sensitive to those kinds of methods that bring this richness of experience to the fore.

Responding to these challenges Callon *et al.* (2009) develop a conception of the social and political dynamics that are often at play in risk regulation and decision-making. For Callon *et al.* that such events and controversies are characterised by often intense public concerns, fuelled by underlying suspicion and lack of trust, and are coupled with high-degrees of scientific uncertainty, suggests that risk issues have the capacity to engender particular kinds of public fora.[4] They term these spaces 'hybrid forums', suggesting that:

> The controversies take place in public spaces that we propose to call hybrid forums – forums because they are open spaces where groups can come together to discuss technical options involving the collective, hybrid because the groups involved and the spokespersons claiming to represent them are heterogeneous, including experts, politicians, technicians and laypersons who consider themselves involved. They are also hybrid because the questions and problems taken up are addressed at different levels in a variety of domains, from ethics to economic and including physiology, nuclear physics, and electromagnetism. (p. 18)

The significance of this concept of hybridity of the political spaces that emerge around particular risk issues is that it captures both the conceptual and practical challenge for critical risk research. For example, Callon *et al.* (2009) outline the nature of this challenge in the following terms:

> Hybrid forums take part in a challenge, a partial challenge at least, to the two great typical divisions of our Western societies: the division that separates specialists and laypersons and the division that distances ordinary citizens from their institutional representatives. These distinctions, and the asymmetries they entail, are scrambled in hybrid forums. Laypersons dare to intervene in technical questions; citizens regroup in order to work out and express new identities, abandoning their usual spokespersons. Thanks to this double transgression, as yet unidentified overflows are revealed and made manageable. The hybrid forums could thus become an apparatus of elucidation. The cost of accepting their use is acceptance of the challenge to the two great divisions. Actors involved in socio-technical controversies are not mistaken. When they establish a new hybrid form, they lay their cards on the table: "We do not accept the monopoly of experts! We want to be directly involved in the political debate on questions that our representatives either ignore or deal with without speaking with us!" (p. 35)

For Callon *et al.* such knowledge controversies represent an opportunity for both engaged research and for realigning the relationships between authorised and non-authorised forms of knowledge. The challenge is to

---

[4] Callon *et al.* conception of the agency of risk issues to draw together a hybrid forum of actors is an extension of recent work on 'issue-oriented' analysis of public controversies. Drawing on pragmatist political and social theory this work suggests that does not preexist the emergence of political issues. Rather public fora are defined around controversies as they emerge, defining the range of actors involved in constituting issues as issues (Latour, 2005; Marres, 2007).

work *with* rather than *against* the heterogeneity invoked in responses to risk problems. If the chapters in this volume offer some, at least partial insights, on this challenge it is to suggest that diverse modes of participation may help, as a means of bridging the gap between those who are studying and those who are being studied; as a means of making those who conduct critical research risk, as well as those who analyse and manage it more accountable; as a means of diversifying the centres of knowledge in contemporary critical risk research.

References

Adam, B., Beck, U., and van Loon, J., eds. (2000) *The Risk Society and Beyond: Critical Issues for Social Theory*. London: Sage.

Amoore, L., and de Goede, M. (2005) Governance, risk and dataveillance in the war on terror. *Crime, Law & Social Change* **43**: 149–173.

———, eds. (2008) *Risk and the War on Terror*. Abingdon, Oxon: Routledge.

Anderson, B. (2010) Security and the future: Anticipating the event of terror. *Geoforum* **41**: 227–235.

Barben, D., Fisher, E., Selin, C., and Guston, D. (2008) Anticapatory governance of nanotechnology: foresight, engagement and integration. In E. J Hackett, O Amsterdamska, M Lynch and J Wajcman (eds.) *The Handbook of Science and Technology Studies – Third Edition*. Cambridge, MA: MIT Press, pp. 979–1000.

Beck, U. (1998) Politics of risk society. In J. Franklin (ed.) *The Politics of Risk Society*. Cambridge: Polity Press, pp. 9–22.

Borraz, O. (2007) Risk and public problems. *Journal of Risk Research* **10**(7): 941–957.

Bracken. L. J., and Oughton, E. A. (eds) (2009) *Interdisciplinarity: Framing, Doing and Application*. Special section of *Area*, **41**(4): 371–481.

Brown, N., and Michael, M. (2002) From authority to authenticity: the changing governance of biotechnology. *Health, Risk and Society* **4**(3): 259–272.

Callon, M., Lascoumes, P., and Barthe, Y. (2009) *Acting in an Uncertain World: An Essay on Technical Democracy*. Cambridge, MA: The MIT Press.

Chilvers, J. (2008) Environmental risk, uncertainty, and participation: mapping an emergent epistemic community. *Environment and Planning A* **40**(12): 2990–3008.

——— (2010) *Sustainable Participation? Mapping out and reflecting on the field of public dialogue on science and technology*. Harwell: Sciencewise Expert Resource Centre.

Collier, S. J. (2008) Enacting catastrophe: preparedness, insurance, budgetary rationalization. *Economy and Society* **37**(2): 224–250.

Colls, R., and Evans, B. (2008) Embodying responsibility: children's health and supermarket initiatives. *Environment & Planning A* **40**: 615–631.

Cooper, M. (2010) Turbulent worlds: financial markets and environmental crisis. *Theory, Culture & Society* **27**(2–3): 167–190.

Dean, M. (1999) Risk, calculable and incalculable. In D. Lupton (ed.) *Risk and Sociocultural Theory*. Cambridge: Cambridge University Press, pp. 131–159.

Diprose, R., Stephenson, N., Mills, C., Race, K., and Hawkins, G. (2008) Governing the future: the paradigm of prudence in political technologies of risk management, *Security Dialogue* **39**(2–3): 267–288.

Douglas, M., and Wildavsky, A. (1982) *Risk and Culture: An Essay on the Selection of Technical and Environmental Dangers*. Berkeley: University of California Press.

Ericson, R. (2005) Governing through risk and uncertainty. *Economy and Society* **34**(4): 659–672.

Fischer, F. (1999) *Citizens, Experts, and the Environment: the Politics of Local Knowledge*. Durham, NC: Duke University Press.

Graham, S., and Wood, D. (2003) Digitizing survelliance: categorisation, space, inequality. *Critical Social Policy* **23**(2): 227–248.

Gusfield, J. R. (1981) *The Culture of Public Problems: Drink-Driving and the Symbolic Order*. Chicago: University of Chicago Press.

Haggerty, K. D. and Ericson, R. V. (2000) The surveillant assemblage. *British Journal of Sociology* **51**(4): 605–622.

Hagendijk, R. P., and Irwin, A. (2006) Public deliberation and governance: engaging with science and technology in contemporary Europe. *Minerva* **44**: 167–184.

Hajer, M. A. (1995) *The Politics of Environmental Discourse: Ecological Modernization and the Policy Process*. Oxford: Oxford University Press.

Jasanoff, S. (1986) *Risk Management and Political Culture: A Comparative Study of Science in the Policy*. New York: Russell Sage Foundation.

Lakoff, A. and Collier, S. J., eds. (2008) *Biosecurity Interventions: Global Health and Security in Question*. New York: Columbia University Press.

Landström, C., Whatmore, S. J., Lane, S. N., Odoni, N. A., Ward, N., and Bradley, S. (2011) Coproducing flood risk knowledge: redistributing expertise in critical 'participatory modelling. *Environment & Planning A* **43**(7): 1617–1633.

Lane, S. N., Odoni, N., Landstroem, C., Whatmore, S. J., Ward, N., and Bradley, S. (2011) Doing flood risk science differently: an experiment in radical scientific method. *Transactions of the Institute of British Geographers* **36**(1): 15–36.

Latour, B. (2005) From realpolitik to dingpolitik or how to make things public. In B. Latour and P. Weibel (eds.) *Making Things Public: Atmospheres of Democracy*. Cambridge, MA: The MIT Press, pp. 4–31.

Lyon, D. (2007) *Surveillance Studies. An Overview*, Polity Press, Cambridge.

Marres, N. (2007) The issues deserve more credit: pragmatist contributions to the study of public involvement in controversy. *Social Studies of Science* **37**(5): 759–780.

Mol, A. (2002) *The Body Multiple: Ontology in Medical Practice*. Durham, NC: Duke University Press.

O'Malley, P. (2004) *Risk, Uncertainty and Government*. London: Glasshouse.

——— (2010) Simulated justice: risk, money and telemetric policing. *British Journal of Crimonology* **50**: 795–807.

Rose, N., and Miller, P. (1992) Political power beyond the State: problematics of government. *British Journal of Sociology* **43**(2): 173–205.

Royal Commission on Environmental Pollution (2008) *Novel Materials in the Environment: The Case of Nanotechnology*. London: HMSO.

Shamir, R. (2010) Capitalism, governance and authority: the case of corporate social responsibility. *Annual Review of Law and Social Science* **6**: 531–553.

Stirling, A. (2009) Risk, uncertainty and power. *Seminar*, **597**, 33–39.

Strathern, M. (2005) Robust knowledge and fragile futures. In A. Ong and S. Collier (eds.) *Global Assemblages: Technology, Politics and Ethics as Anthropological Problems*. Oxford: Blackwell, pp. 464–481.

Verran, H. (2001) *Science and an African Logic*. Chicago: University of Chicago Press.

Winner, L. (1978) *Autonomous Technology: Technics-out-of-Control as a Theme in Political Thought*. Cambridge, MA: MIT Press.

Wynne, B. (1996) May the sheep safely graze? A reflexive view of the expert-lay knowledge divide. In S. M. Lash, B. Szerszynski and B. Wynne (eds.) *Risk, Environment and Modernity: Towards a New Ecology*. London: Sage, pp. 44–83.

——— (2001) Creating public alienation: expert cultures of risk and ethics on GMOs. *Science as Culture* **10**(3): 445–481.

——— (2006) Public engagement as a means of restoring public trust in science – hitting the notes, but missing the music? *Community Genetics* **9**(3): 211–220.

Index

Critical Risk Research: Practices, Politics and Ethics, First Edition.
Edited by Matthew Kearnes, Francisco Klauser and Stuart Lane.